Vanua

T0342676

MATHS 5 PLUS

Harry O'Brien   Greg Purcell

OXFORD
UNIVERSITY PRESS
AUSTRALIA & NEW ZEALAND

Contents

Contents

Thinking and Working through Investigations

The *Maths Plus* Vanuatu Edition is a whole-school numeracy program based on the Mathematics 1–10 Syllabus and provides a sound foundation for the teaching and learning of mathematics. This edition has been revised to address new Essential Learnings.

The student books develop mathematical knowledge and understanding and students apply these skills to undertake various mathematical tasks. The activities allow students to think and reason mathematically whilst building up their knowledge and understanding of essential mathematical concepts. The diagnostic review pages help pinpoint the strengths and weaknesses of each student, thus providing opportunities for specific re-teaching. The 16 investigations in each book allow students to apply their knowledge and skills to undertake relevant and stimulating investigative activities.

Example

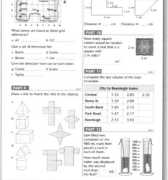

Concept and skills development

Assess concepts and skills

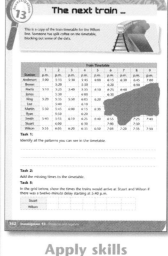

Apply skills

A 'My investigation review' page is included in each student book. This page is photocopiable and can be given out to students before they commence an investigation. Students will be able to record succinctly how they will go about completing their investigations, what strategies they will employ and any conclusions they might reach. The completed review page will provide added evidence about the processes individual students use in undertaking investigative activities.

Find a topic

Find a Level 3 outcome

Find a Level 3 outcome

Making to decades

1 Add these numbers by adding the tens, then making to a decade.

37 + 27 = ?
Think
37 + 20 + 3 + 4.
37 ... 57 ... 60 ... 64.

a 35 + 16 becomes | 35 | + | 10 | + | 5 | + | 1 | = | 51 |

b 67 + 25 becomes [ ] + [ ] + [ ] + [ ] = [ ]

c 38 + 46 becomes [ ] + [ ] + [ ] + [ ] = [ ]

d 46 + 48 becomes [ ] + [ ] + [ ] + [ ] = [ ]

2 Add these larger numbers using the 'making to decades' strategy.

a 127 + 26 becomes [ ] + [ ] + [ ] + [ ] = [ ]

b 236 + 37 becomes [ ] + [ ] + [ ] + [ ] = [ ]

138 + 26 = ?
Think
138 + 20 + 2 + 4.

c 256 + 28 becomes [ ] + [ ] + [ ] + [ ] = [ ]

d 328 + 35 becomes [ ] + [ ] + [ ] + [ ] = [ ]

e 427 + 46 becomes [ ] + [ ] + [ ] + [ ] = [ ]

f 315 + 37 becomes [ ] + [ ] + [ ] + [ ] = [ ]

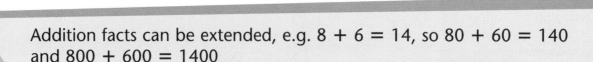

Addition facts can be extended, e.g. 8 + 6 = 14, so 80 + 60 = 140 and 800 + 600 = 1400

3 Complete these examples to extend the basic number facts, then check them with a calculator.

7 + 5 = 12
7 tens + 5 tens
makes 12 tens
so 70 + 50 = 120.

a | 70 | + | 60 | = | 130 | | 700 | + | 600 | = | 1300 |

b | 50 | + | 40 | = [ ] | 500 | + | 400 | = [ ]

c | 60 | + | 50 | = [ ] | 600 | + | 500 | = [ ]

d | 160 | + | 50 | = [ ] | 1600 | + | 500 | = [ ]

e | 500 | + | 70 | = [ ] | 5000 | + | 700 | = [ ] | 5000 | + | 7000 | = [ ]

f | 900 | − | 40 | = [ ] | 9000 | − | 400 | = [ ] | 9000 | − | 4000 | = [ ]

g | 800 | − | 60 | = [ ] | 8000 | − | 600 | = [ ] | 8000 | − | 6000 | = [ ]

N3.2 Students identify and solve addition and subtraction problems involving whole numbers, and decimal fractions in context, selecting from a range of computation methods, strategies and known number facts.

Place value

4 Write how many pieces of each MAB material would be needed to make the numbers, then record the numbers on the number expanders.

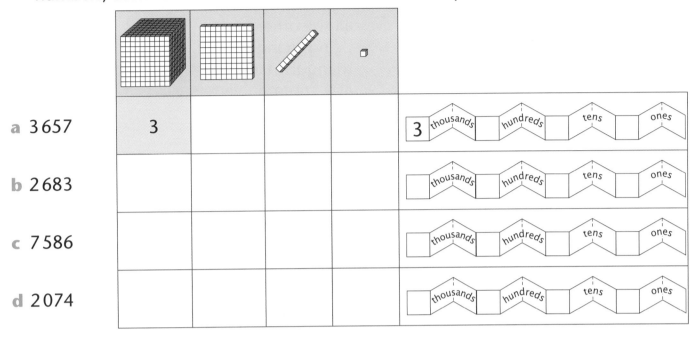

a 3 657	3				3 thousands hundreds tens ones
b 2 683					thousands hundreds tens ones
c 7 586					thousands hundreds tens ones
d 2 074					thousands hundreds tens ones

5 Write the place value of each bold number.   **thousands, hundreds, tens, ones**

a 4**7**53 ___hundreds___   **d** 3**5**42 _____   **g** 673**9** _____

b 2**5**74 _____   **e** **7**283 _____   **h** 6**4**16 _____

c 63**5**1 _____   **f** **5**434 _____   **i** 345**6** _____

6 Write the number before and after the shaded number on each line.

	Before		After			Before		After
a		564			**d**		5 462	
b		387			**e**		6 999	
c		400			**f**		3 001	

7 Order the following numbers from smallest to largest.

a 46, 247, 56, 474   _____

b 357, 323, 531, 784   _____

c 2 374, 7 423, 3 724, 2 743   _____

d 2 701, 2 671, 2 761, 3 017   _____

e 8 603, 3 806, 6 803, 6 380   _____

Three-dimensional shapes

Prisms have two faces called bases that are parallel and congruent. All the other faces on a prism are rectangles.

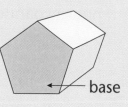

base

Pyramids have only one base with all the other faces being triangles. The triangular faces meet at a common vertex.

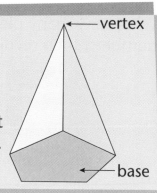

vertex

base

8 Place the letters in the correct position on the grid to identify the cylinders, cones, spheres, prisms and pyramids.

a

b

c

d

e

f

g

h

i

j

k

Cylinders				
Cones				
Spheres				
Prisms				
Pyramids				

l

m

9 Find two items in your school that are:

a prisms

b cylinders

c cones

d spheres

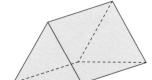

10 Explain why this object is a prism.

_____

S3.1 Students describe the defining geometric properties of families of 3D shapes, model 3D shapes using nets and other representations, and identify and describe the properties of specific families and subgroups of 2D shapes.

Length can be measured in centimetres. The prefix **centi** means 100.
A centimetre is one-hundredth of a metre.

11 Use a ruler to measure the length of these lines in centimetres.

a _____ _____ cm

b _____ _____ cm

c _____ _____ cm

d _____ _____ cm

e _____ _____ cm

f _____ _____ cm

12 Stand a metre rule against your side to see how long a metre is.

13 Lay a metre rule on the ground and see if you can step out 1 metre.

0 cm 10 cm 20 cm 30 cm 40 cm 50 cm 60 cm 70 cm 80 cm 90 cm 100 cm

14 Use the diagram of the metre rule above to complete these activities.

a Write $\frac{1}{2}$ m on the ruler half way along the ruler.

b How many centimetres are there in half a metre? _____

c Write $\frac{1}{4}$ m on the ruler a quarter of the way along the ruler.

d How many centimetres are in a quarter of a metre? _____

e Write 1 m at the end ruler.

f How many centimetres in one metre? _____

g Write $\frac{3}{4}$ m on the ruler three-quarters of the way along the ruler.

h How many centimetres in three-quarters of a metre? _____

15 Estimate and measure the length of each pencil.

	Estimate	Centimetres
a		
b		
c		
d		

M3.1 Students identify and use equivalent forms of standard units when measuring, comparing and ordering, and estimate using a range of personal referents.

5

Subtraction strategies

1 Take away the tens part then the ones part to solve the subtraction.

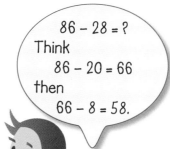

86 − 28 = ?
Think
    86 − 20 = 66
then
    66 − 8 = 58.

a 86 − 24 =

b 45 − 23 =

c 59 − 34 =

d 86 − 35 =

e 45 − 27 =

f 56 − 34 =

g 74 − 45 =

h 83 − 56 =

i 92 − 63 =

j 94 − 62 =

2 Complete these by extending the subtraction facts.

a 7 − 5 = 2   70 − 50 = 20   700 − 500 = 200

b 8 − 4 =   80 − 40 =   800 − 400 =

c 9 − 6 =   90 − 60 =   900 − 600 =

d 8 − 3 =   80 − 30 =   800 − 300 =

e 9 − 5 =   90 − 50 =   900 − 500 =

f 6 − 4 =   60 − 40 =   600 − 400 =

g 9 − 8 =   90 − 80 =   900 − 800 =

3 Count back to solve the subtractions.

168 − 30
Count back 30.
158 ... 148 ... 138.

a 58 − 10 =

b 62 − 20 =

c 98 − 30 =

d 158 − 20 =

e 279 − 30 =

f 546 − 10 =

g 693 − 100 =

h 785 − 300 =

i 419 − 200 =

j 806 − 100 =

k 777 − 200 =

l 591 − 300 =

N3.2 Students identify and solve addition and subtraction problems involving whole numbers, and decimal fractions in context, selecting from a range of computation methods, strategies and known number facts.

Halves, quarters and eighths

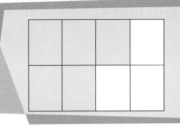

$\dfrac{5}{8}$ —— The **numerator** shows us how many equal parts. (The fractional part.)

—— The **denominator** shows us how many equal parts are in the whole.

4 Shade the fraction of each shape.

a 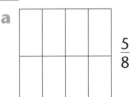 $\dfrac{5}{8}$

b  $\dfrac{3}{4}$

c 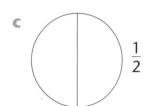 $\dfrac{1}{2}$

d 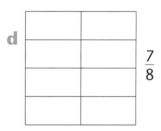 $\dfrac{7}{8}$

5 Alexa has coloured $\dfrac{1}{8}$ of her strip of paper. Colour the other strips to match the labels.

a | $\dfrac{1}{8}$

b | $\dfrac{1}{2}$

c | $\dfrac{1}{4}$

d | $\dfrac{3}{4}$

$\dfrac{1}{4}$ is twice as much as $\dfrac{1}{8}$.

6 Shade the fraction of each group.

a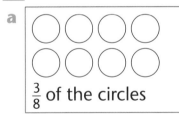

$\dfrac{3}{8}$ of the circles

b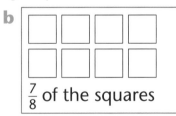

$\dfrac{7}{8}$ of the squares

c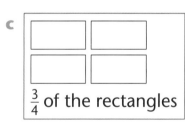

$\dfrac{3}{4}$ of the rectangles

7 Label the halves, quarters and eighths on the number lines.

a 0 $\boxed{\dfrac{}{2}}$ 1 halves

b 0 $\boxed{\dfrac{}{4}}$ $\boxed{\dfrac{}{4}}$ $\boxed{\dfrac{}{4}}$ 1 quarters

c 0 $\boxed{\dfrac{}{8}}$ $\boxed{\dfrac{}{8}}$ $\boxed{\dfrac{}{8}}$ $\boxed{\dfrac{}{8}}$ $\boxed{\dfrac{}{8}}$ $\boxed{\dfrac{}{8}}$ $\boxed{\dfrac{}{8}}$ 1 eighths

8 Order the fractions from smallest to largest.

a $\dfrac{7}{8}$ $\dfrac{1}{8}$ $\dfrac{5}{8}$ $\dfrac{3}{8}$

b $\dfrac{1}{4}$ $\dfrac{3}{4}$ $\dfrac{2}{4}$ $\dfrac{1}{8}$

N3.1 Students compare, order and represent whole numbers to 9 999 and common and decimal fractions, calculate cash transactions and describe other methods of payment.

7

9 A group of students in 5B were given a 10-word spelling quiz. They then made a graph to show their scores.

a Who scored the highest? _____

b Who scored the lowest? _____

c Who scored six less than Jimmy?

_____

d How many more did Aimee score than Zlatco? _____

e Whose scores were the same?

_____

f Who scored four fewer than Jimmy?

_____

g Who scored four more than Angel?

_____

h Whose score was in the middle?

_____

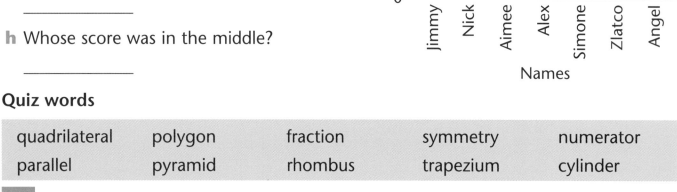

Spelling quiz scores

Number of times with no mistakes

Names

Jimmy · Nick · Aimee · Alex · Simone · Zlatco · Angel

Quiz words

quadrilateral	polygon	fraction	symmetry	numerator
parallel	pyramid	rhombus	trapezium	cylinder

10 Select a group of 7 people to spell the quiz words in the box above. Record their names in a table and then make a bar graph to show their results.

Name	Score

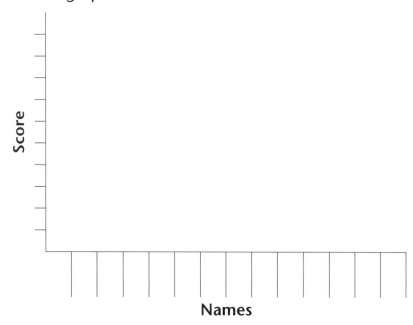

Score

Names

CD3.2 Students design and trial a variety of data collection methods and use existing sources of data to investigate their own and others' questions, organise data and create suitable displays identifying and interpreting elements of the displays.

11 Estimate and then measure the length of the arrows in centimetres.

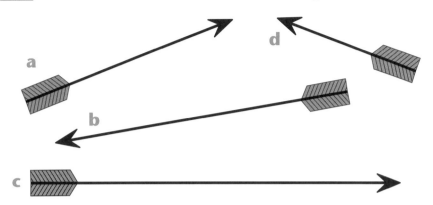

	Estimate	Actual
a	cm	cm
b	cm	cm
c	cm	cm
d	cm	cm

12 Use the knowledge you gained above to estimate and then to measure the lengths of the following items.

	Length	Estimate	Centimetres
a	Your pencil		
b	A book		
c	A calculator		
d	A glue stick		
e	Your shoe		
f	Your desk		

13 With chalk mark a 10-m length on the playground then count how many of your normal steps are taken to cover the distance.

14 Use this information to estimate and mark these lengths on the playground using chalk. Check how close your estimate is by measuring out the exact length. Tick the 'Close' box if you were happy with your estimate.

	Length	Steps	Close
a	5 metres		
b	10 metres		
c	20 metres		
d	25 metres		

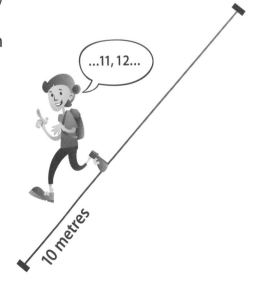

...11, 12...

10 metres

M3.1 Students identify and use equivalent forms of standard units when measuring, comparing and ordering, and estimate using a range of personal referents.

9

Revising multiplication facts

1 Solve the mutiplication facts using the arrays if you need them.

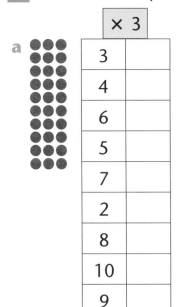

a

× 3	
3	
4	
6	
5	
7	
2	
8	
10	
9	

b

× 4	
3	
4	
6	
5	
7	
2	
8	
10	
9	

c

× 6	
3	
4	
6	
5	
7	
2	
8	
10	
9	

POTATOES $2 kg

MUSHROOMS $6 kg

TOMATOES $4 kg

CABBAGE $3 ea.

BEANS $2 kg

2 Calculate the cost of each family's shopping.

a **The Younis family**

Item	Cost
10 kg of potatoes	
3 kg of mushrooms	
2 cabbages	
6 kg of beans	
4 kg of tomatoes	
Total	

b **The Walters family**

Item	Cost
5 kg of potatoes	
2 kg of mushrooms	
3 cabbages	
7 kg of beans	
3 kg of tomatoes	
Total	

3 Write a multiplication problem based on the items above.

N3.3 Students identify and solve multiplication and division problems involving whole numbers, and decimal fractions in context, selecting from a range of computation methods, strategies and known number facts.

Addition and multiplication turnarounds

4 George said that it doesn't matter in which order numbers are added. Check to see if George is correct by adding the following pairs of number sentences.

70 + 90 = 160
90 + 70 = 160
Cool!

a 40 + 60 = ⬚   60 + 40 = ⬚

b 80 + 30 = ⬚   30 + 80 = ⬚

c 70 + 50 = ⬚   50 + 70 = ⬚

d 95 + 25 = ⬚   25 + 95 = ⬚

e 76 + 34 = ⬚   34 + 76 = ⬚

f 58 + 60 = ⬚   60 + 58 = ⬚

5 Maria said that you can also multiply numbers in any order. Check to see if Maria is correct by multiplying the following pairs of numbers.

a 60 × 3 = ⬚   3 × 60 = ⬚

b 50 × 4 = ⬚   4 × 50 = ⬚

c 60 × 5 = ⬚   5 × 60 = ⬚

d 80 × 3 = ⬚   3 × 80 = ⬚

e 85 × 2 = ⬚   2 × 85 = ⬚

f 65 × 4 = ⬚   4 × 65 = ⬚

g 88 × 6 = ⬚   6 × 88 = ⬚

So is 50 x 7
equal to 7 x 50?

h Do you agree that numbers can be multiplied in any order? _____

6 Write two number sentences for each problem before solving them.

	Problem	Number sentences	Answer
a	John had 135 marbles and Jim had 125. How many marbles did they have altogether?	⬚ + ⬚ = ⬚ ⬚ + ⬚ = ⬚	
b	Six rows of flowers were planted in a garden. How many flowers were there if there were 20 in each row?	⬚ × ⬚ = ⬚ ⬚ × ⬚ = ⬚	

N3.2 Students identify and solve addition and subtraction problems involving whole numbers, and decimal fractions in context, selecting from a range of computation methods, strategies and known number facts.

11

Reading a map

7 Study the map in order to answer the questions.

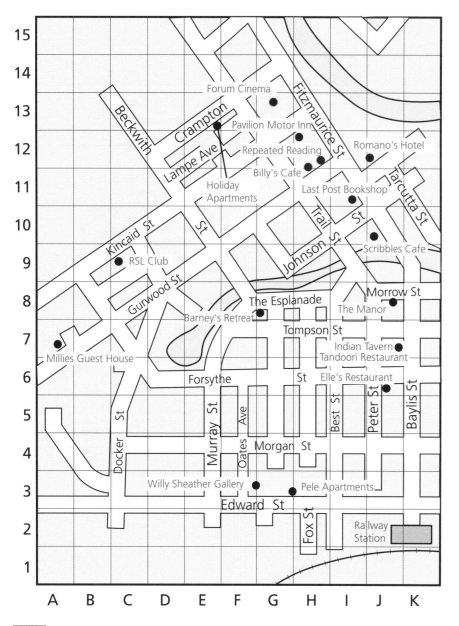

Give the grid references for these points on the map.
a The RSL club ____
b Millies Guest House ____
c Elle's Restaurant ____
d Forum Cinema ____
e Scribbles Cafe ____
f The Manor ____

8 What can be found at these grid references?
a G8 _____
b I11 _____
c K2 _____

9 Add a swimming pool at A15, B15, C15, A14, B14, C14.

10 Describe one way of getting from the Railway Station to Barney's Retreat.

_____

_____

_____

11 Where am I?

• I left the railway station and turned left into Edward St.
• I drove along Edward St and turned right into Docker St.
• I drove along Docker St until I got to Gurwood St, then I turned left.
• I parked my car in Kincaid St.

S3.2 Students interpret and create maps and plans using a range of conventions, describe locations and give directions using major compass points, angles and grids.

Perimeter

Perimeter is the distance around the outside of a shape.
(The length of its boundary.)

12 Measure the perimeter of these shapes.

a

Perimeter
= ____ cm

b

Perimeter
= ____ cm

c

Perimeter = ____ cm

d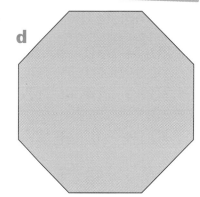

Perimeter = ____ cm

13 Use the 1-cm grid paper to construct squares of the given perimeters.
Starting points have been given.

a   b   c   d

8 cm   12 cm   16 cm   20 cm

14 Use the 1-cm dot paper to calculate the perimeter of each shape.

a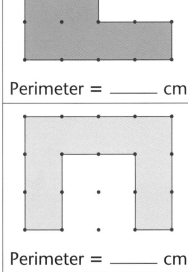

Perimeter = _____ cm

b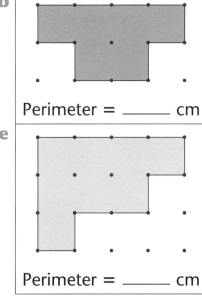

Perimeter = _____ cm

c

Perimeter = _____ cm

d

Perimeter = _____ cm

e

Perimeter = _____ cm

f

Perimeter = _____ cm

M3.1 Students identify and use equivalent forms of standard units when measuring,
comparing and ordering, and estimate using a range of personal referents.

13

When numbers are added it doesn't matter which order they are added in. E.g. 3 + 4 + 6 = 13 and 6 + 4 + 3 = 13.

70 + 90 + 30 is equal to 70 + 30 + 90.

1 Complete the sets of number sentences to see if this is true.

a 20 + 70 + 80 = ☐   20 + 80 + 70 = ☐

b 30 + 14 + 120 = ☐   30 + 120 + 14 = ☐

c 30 + 8 + 270 = ☐   270 + 30 + 8 = ☐

d 180 + 56 + 120 = ☐   180 + 120 + 56 = ☐

e 68 + 49 + 32 = ☐   68 + 32 + 49 = ☐

When numbers are multiplied it doesn't matter which order they are multiplied in. E.g. 3 × 4 × 5 = 60 and 5 × 4 × 3 = 60.

2 Complete the sets of number sentences to see if this is true.

7 × 2 × 3 equals 2 × 7 × 3.

a 2 × 3 × 4 = ☐   4 × 2 × 3 = ☐

b 4 × 3 × 1 = ☐   3 × 1 × 4 = ☐

c 5 × 4 × 2 = ☐   2 × 5 × 4 = ☐

d 3 × 4 × 5 = ☐   5 × 4 × 3 = ☐

e 2 × 5 × 6 = ☐   6 × 5 × 2 = ☐

3 Rewrite each number sentence to make it easier to solve.

65 + 83 + 35 becomes 65 + 35 + 83.

a 60 + 47 + 40 becomes ☐ + ☐ + ☐ = ☐

b 75 + 23 + 25 becomes ☐ + ☐ + ☐ = ☐

c 70 + 90 + 30 becomes ☐ + ☐ + ☐ = ☐

d 180 + 38 + 20 becomes ☐ + ☐ + ☐ = ☐

e 7 × 2 × 5 becomes ☐ × ☐ × ☐ = ☐

f 7 × 4 × 5 becomes ☐ × ☐ × ☐ = ☐

N3.2 Students identify and solve addition and subtraction problems involving whole numbers, and decimal fractions in context, selecting from a range of computation methods, strategies and known number facts.   N3.3

Fifths and tenths

4 Colour the containers to a level to match the labels which show how full they are.

a	b	c	d	e	f
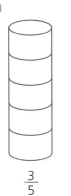	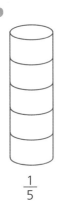	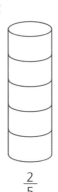	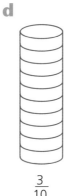	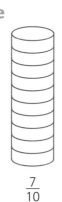	
$\frac{3}{5}$	$\frac{1}{5}$	$\frac{2}{5}$	$\frac{3}{10}$	$\frac{7}{10}$	$\frac{9}{10}$

5 Circle the fractions of each collection.

a	b	c
$\frac{3}{5}$ of the balls	$\frac{7}{10}$ of the flowers	$\frac{3}{10}$ of the egg cups

6 Label the halves, fifths and tenths on the number lines.

a   0 $\frac{}{2}$ 1   halves

b   0 $\frac{}{5}$ $\frac{}{5}$ $\frac{}{5}$ $\frac{}{5}$ 1   fifths

c   0 $\frac{}{10}$ $\frac{}{10}$ $\frac{}{10}$ $\frac{}{10}$ $\frac{}{10}$ $\frac{}{10}$ $\frac{}{10}$ $\frac{}{10}$ $\frac{}{10}$ 1   tenths

7 Use the number line to help you order the fractions from smallest to largest.

a	$\frac{3}{10}$	$\frac{7}{10}$	$\frac{5}{10}$	$\frac{4}{10}$		d	$\frac{4}{5}$	$\frac{1}{5}$	$\frac{2}{5}$	$\frac{3}{5}$	
b	$\frac{1}{10}$	$\frac{4}{10}$	$\frac{3}{10}$	$\frac{3}{5}$		**e**	$\frac{2}{5}$	$\frac{5}{5}$	$\frac{1}{2}$	$\frac{4}{5}$	
c	$\frac{9}{10}$	$\frac{1}{2}$	$\frac{3}{10}$	$\frac{8}{10}$		**f**	$\frac{1}{2}$	$\frac{1}{10}$	$\frac{1}{5}$	$\frac{7}{10}$	

N3.1 Students compare, order and represent whole numbers to 9 999 and common and decimal fractions, calculate cash transactions and describe other methods of payment.

15

Quadrilaterals

Quadrilaterals are shapes that have 4 straight sides. There are many types of quadrilaterals and some have special names.

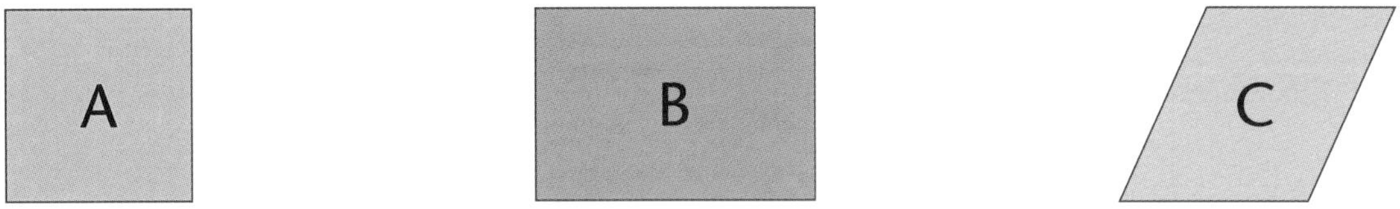

A

B

C

A **square** is a quadrilateral that has 4 equal sides and 4 right angles.	A **rectangle** is a quadrilateral that has 2 pairs of equal sides and 4 right angles.	A **rhombus** is a quadrilateral that has 4 equal sides and the opposite angles are equal.

D

E

F

A **trapezium** is a quadrilateral with one set of parallel sides.	A **parallelogram** is any quadrilateral with 2 sets of parallel sides.	4-sided shapes with no special name are given the family name of **quadrilaterals**.

8 Draw an example of each quadrilateral above.

a

b

c

d

e

f

9 Answer the questions.

a Do all quadrilaterals have parallel sides? _____

b Is a triangle a quadrilateral? _____

c How many sides would there be on 8 quadrilaterals? _____

S3.1 Students describe the defining geometric properties of families of 3D shapes, model 3D shapes using nets and other representations, and identify and describe the properties of specific families and subgroups of 2D shapes.

How tall are students in Year 5?

On this page you will compare the heights of six students in class 5B at another school with six students in your class.

10 Use the information in the table to create a bar graph showing the heights of the class.

Names	Height
Sue	131 cm
Jon	126 cm
Sam	140 cm
Kia	139 cm
Peg	124 cm
Ali	136 cm

Class heights of 5B

Centimetres

140
139
138
137
136
135
134
133
132
131
130
129
128
127
126
125
124
123
122
121
120

Sue   Jon   Sam   Kia   Peg   Ali

Names

11 Measure and graph the height of six people in your class.

12 Compare the heights of the students in your class with those in 5B and explain what you noticed.

_____
_____
_____
_____
_____
_____
_____

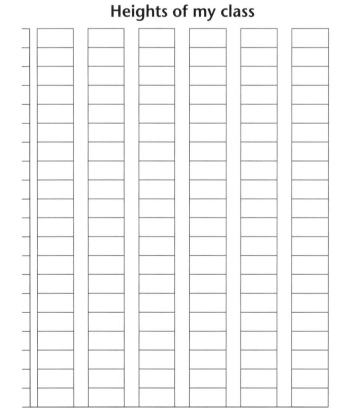

Heights of my class

Centimetres

Names

CD3.2 Students design and trial a variety of data collection methods and use existing sources of data to investigate their own and others' questions, organise data and create suitable displays identifying and interpreting elements of the displays.

17

- 5 ones plus 8 ones equals 13 ones. Exchange 10 ones for 1 ten then record the 3 in the ones column.

Hund	Tens	Ones
1	1	
2	7	5
+ 5	6	8
8	4	3

- 1 ten plus 7 tens plus 6 tens equals 14 tens. Exchange 10 tens for 1 hundred then record the 4 in the tens column.
- 1 hundred plus 2 hundreds plus 5 hundreds equals 8 hundreds.

1 Complete the three-digit additions (trading in the ones and tens).

a

Hund	Tens	Ones
1	6	4
+ 2	4	8

b

Hund	Tens	Ones
2	3	4
+ 4	6	3

c

Hund	Tens	Ones
5	4	8
+ 2	1	4

d

Hund	Tens	Ones
1	0	6
+ 6	8	8

e

Hund	Tens	Ones
6	9	6
+	4	5

f

Hund	Tens	Ones
5	7	4
+ 1	4	8

2 Supply the missing numbers for each addition.

a
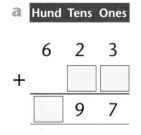

Hund	Tens	Ones
6	2	3
+	☐	☐
☐	9	7

b
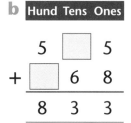

Hund	Tens	Ones
5	☐	5
+ ☐	6	8
8	3	3

c
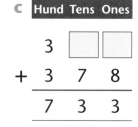

Hund	Tens	Ones
3	☐	☐
+ 3	7	8
7	3	3

d
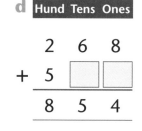

Hund	Tens	Ones
2	6	8
+ 5	☐	☐
8	5	4

3 Use the 'make to a hundred' strategy to solve these additions mentally.

a 285 + 45 becomes | 285 | + | 15 | + | 30 | = | 330 |

285 + 45
Think.
285 + 15 = 300
300 + 30 = 330

b 496 + 34 becomes ☐ + ☐ + ☐ = ☐

c 295 + 25 becomes ☐ + ☐ + ☐ = ☐

d 588 + 32 becomes ☐ + ☐ + ☐ = ☐

e 392 + 38 becomes ☐ + ☐ + ☐ = ☐

f 490 + 63 becomes ☐ + ☐ + ☐ = ☐

g 297 + 83 becomes ☐ + ☐ + ☐ = ☐

h 584 + 76 becomes ☐ + ☐ + ☐ = ☐

i 675 + 45 becomes ☐ + ☐ + ☐ = ☐

N3.2 Students identify and solve addition and subtraction problems involving whole numbers, and decimal fractions in context, selecting from a range of computation methods, strategies and known number facts. **PA3.2**

Extending multiplication facts

We know that 4 x 3 = 12 so 4 x 3 tens must equal 12 tens.
E.g. 4 x 30 = 120.

4 Complete the examples to extend the multiplication facts.

5 x 4 = 20
so 5 x 40 = 200.

a	2	×	3	=	6	2	× 30 = 60
b	2	×	4	=		2	× 40 =
c	3	×	3	=		3	× 30 =
d	3	×	5	=		3	× 50 =
e	3	×	8	=		3	× 80 =
f	5	×	3	=		5	× 30 =
g	5	×	5	=		5	× 50 =
h	5	×	6	=		5	× 60 =
i	4	×	3	=		4	× 30 =
j	4	×	6	=		4	× 60 =
k	4	×	8	=		4	× 80 =

Another strategy used to multiply by a multiple of 10 is to use repeated addition. E.g. 4 x 20 = becomes 20 + 20 + 20 + 20 = 80.

5 Use repeated addition to find the answers to these questions.

3 x 90 = 270
90 + 90 + 90 = 270

a 3 × 40 = [ ] + [ ] + [ ] = [ ]
b 3 × 20 = [ ] + [ ] + [ ] = [ ]
c 4 × 50 = [ ] + [ ] + [ ] + [ ] = [ ]
d 4 × 30 = [ ] + [ ] + [ ] + [ ] = [ ]
e 5 × 20 = [ ] + [ ] + [ ] + [ ] + [ ] = [ ]
f 5 × 40 = [ ] + [ ] + [ ] + [ ] + [ ] = [ ]

Remember: **a line of symmetry** creates a mirror image so that both sides of the shape match each other exactly.

6 Decide whether these shapes are symmetrical or asymmetrical, then draw as many lines of symmetry as you can on them. The first one has been done for you.

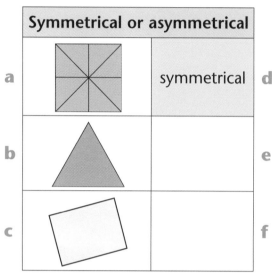

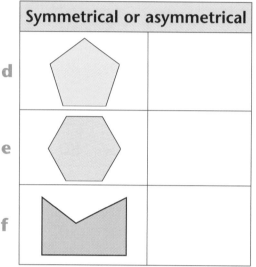

Symmetrical or asymmetrical	
a	symmetrical
b	
c	

Symmetrical or asymmetrical	
d	
e	
f	

A butterfly is **symmetrical**.

Shapes that are not symmetrical are called **asymmetrical**.

7 Draw the other half of each shape by using its line of symmetry as a starting point.

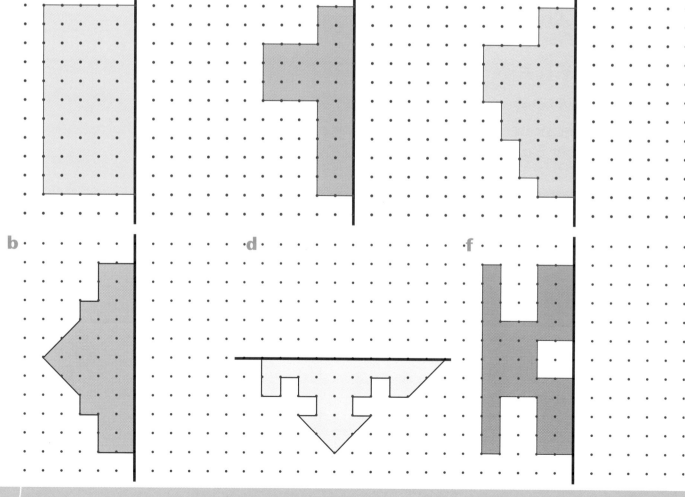

a   c   e

b   d   f

S3.1 Students describe the defining geometric properties of families of 3D shapes, model 3D shapes using nets and other representations, and identify and describe the properties of specific families and subgroups of 2D shapes.

The square centimetre

Small areas are measured using **square centimetres**.
The symbol for square centimetres is **cm²**.
Place the tip of your finger inside the square centimetre at
the top of the page to see how close it is to a square centimetre.

1 cm

1 cm

8 Each MAB one or centicube covers an area of 1 square centimetre.
How many MAB ones fit on these areas? Check using your fingertip.

a b d

c

e

a Area = _____   b Area = _____   c Area = _____   d Area = _____   e Area = _____

9 Calculate the area of the shapes by counting the square centimetres.

a

b

c

☐ square cm   ☐ square cm   ☐ square cm

d

e

f

☐ square cm   ☐ square cm   ☐ square cm

g

h

☐ square cm   ☐ square cm

M3.1 Students identify and use equivalent forms of standard units when measuring, comparing and ordering, and estimate using a range of personal referents.

21

Multiplication—the 7's facts

1 Multiplication facts can be built up from other multiplication facts. Build other multiplication facts from the ones given for the 7's facts.

I know 5 × 7 = 35 so 6 × 7 must be 35 + 7.

a $2 \times 7 = 14$ so $3 \times 7 = \boxed{}$

b $4 \times 7 = 28$ so $5 \times 7 = \boxed{}$

c $8 \times 7 = 56$ so $9 \times 7 = \boxed{}$

d $9 \times 7 = 63$ so $10 \times 7 = \boxed{}$

e $10 \times 7 = 70$ so $11 \times 7 = \boxed{}$

2 The 'building down' strategy can also be used to create multiplication facts. Build down to create these multiplication facts for times 7.

9 × 7 = 63 so 8 × 7 is equal to 63 – 7.

a $7 \times 7 = 49$ so $6 \times 7 = \boxed{}$

b $4 \times 7 = 28$ so $3 \times 7 = \boxed{}$

c $8 \times 7 = 56$ so $7 \times 7 = \boxed{}$

When we multiply numbers together we call the result the **product**.

3 Answer the questions.

a Product of 3 and 7 = ☐   f Product of 3 and 3 = ☐

b Product of 2 and 7 = ☐   g Product of 8 and 6 = ☐

c Product of 4 and 7 = ☐   h Product of 8 and 5 = ☐

d Product of 5 and 7 = ☐   i Product of 6 and 7 = ☐

e Product of 6 and 6 = ☐   j Product of 4 and 6 = ☐

The product of 9 and 7 is 63.

4 Write two sets of numbers that could be multiplied to give a product of 24.

☐ × ☐   ☐ × ☐

5 Complete the spider webs.

a

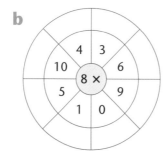

b

c

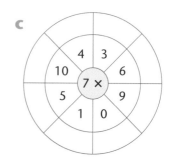

N3.3 Students identify and solve multiplication and division problems involving whole numbers, and decimal fractions in context, selecting from a range of computation methods, strategies and known number facts.

Jump strategy

When we use the **jump strategy**, we add or subtract the tens part to the first number, before adding or subtracting the ones.
The jump strategy can be easily seen on this number line.

$$120 + 35 = 155$$

120 122 124 126 128 130 132 134 136 138 140 142 144 146 148 150 152 154 156 158 160 162
121 123 125 127 129 131 133 135 137 139 141 143 145 147 149 151 153 155 157 159 161 163

6 Use the jump strategy to answer the questions.

257 − 23 = ?
Think 257 − 20 = 237.
Now take away 3,
equals 234.

a | 136 + 36 | think | 136 | + | 30 | + | 6 | = | ☐
b | 245 + 47 | think | ☐ | + | ☐ | + | ☐ | = | ☐
c | 368 − 36 | think | ☐ | − | ☐ | − | ☐ | = | ☐
d | 547 − 34 | think | ☐ | − | ☐ | − | ☐ | = | ☐
e | 626 + 57 | think | ☐ | + | ☐ | + | ☐ | = | ☐

7 Use the jump strategy to solve these questions mentally.

a | 422 | + | 37 | = | ☐
b | 421 | + | 39 | = | ☐
c | 463 | − | 27 | = | ☐
d | 465 | − | 38 | = | ☐
e | 427 | + | 26 | = | ☐

f | 459 | − | 33 | = | ☐
g | 426 | + | 35 | = | ☐
h | 457 | − | 35 | = | ☐
i | 862 | + | 37 | = | ☐
j | 987 | − | 45 | = | ☐

8 Try the jump strategy on these questions involving two 3-digit numbers. You may need the working space.

	Question	Working	Answer
a	362 + 227	362 … 562 … 582 …	
b	456 + 337		
c	238 + 548		

PA3.2   N3.2 Students identify and solve addition and subtraction problems involving whole numbers, and decimal fractions in context, selecting from a range of computation methods, strategies and known number facts.

23

Angles are classified according to the amount of turn between two arms.

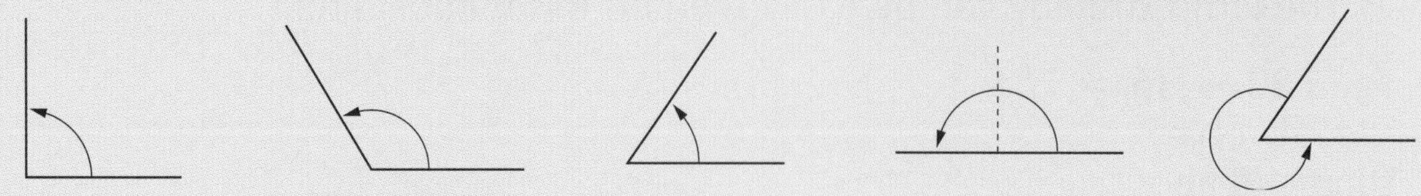

Right angle	Obtuse angle	Acute angle	Straight angle	Reflex angle
Square corner 90°	Greater than 90° but less than 180°	Less than 90°	Can be made from 2 right angles 180°	Greater than 180° but less than 360°

9 Label the angles either right angle, obtuse, acute, reflex or straight.

a

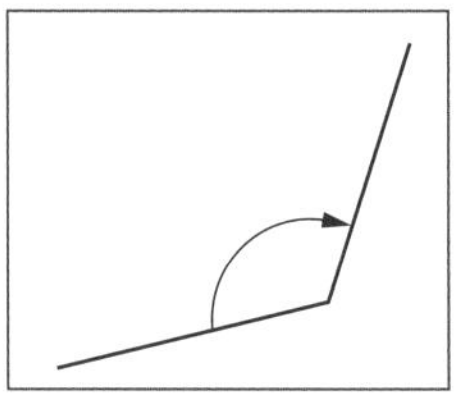

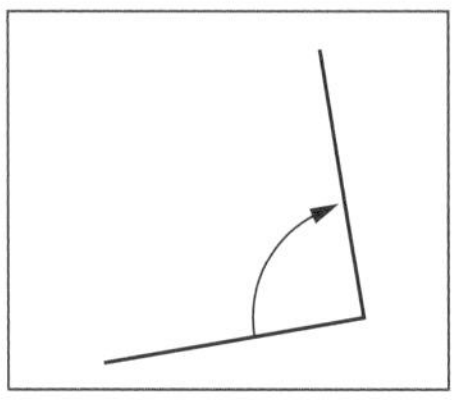

d

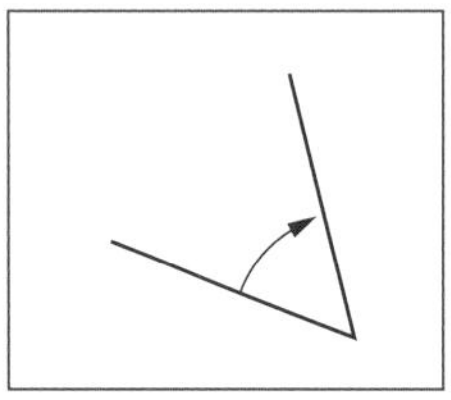

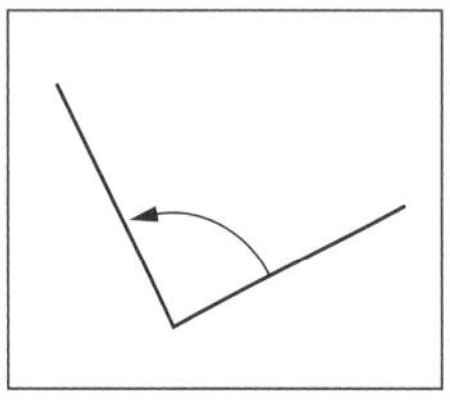

g

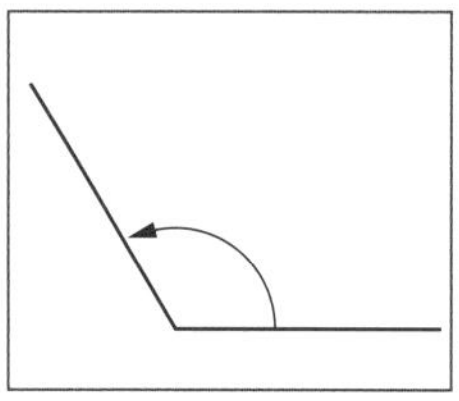

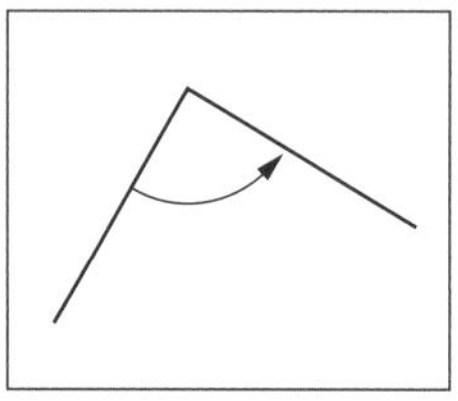

j

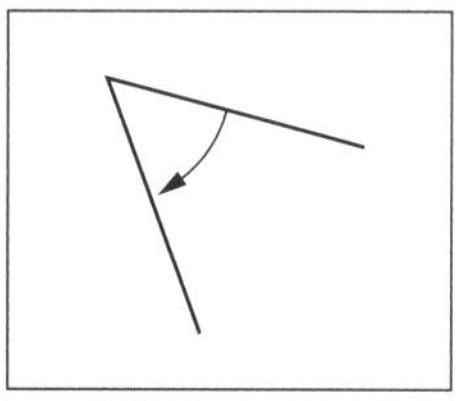

b

e

h

k

c

f

i

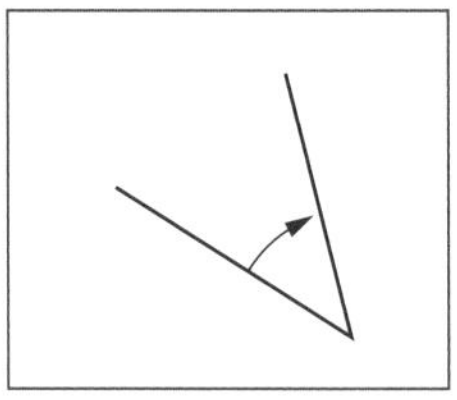

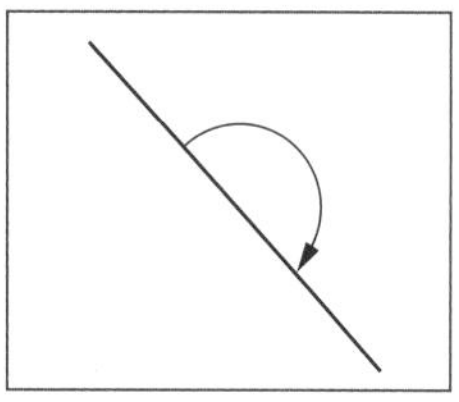

l

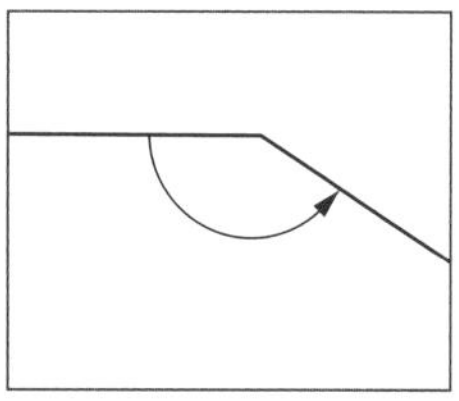

10 Identify which shape is being described.

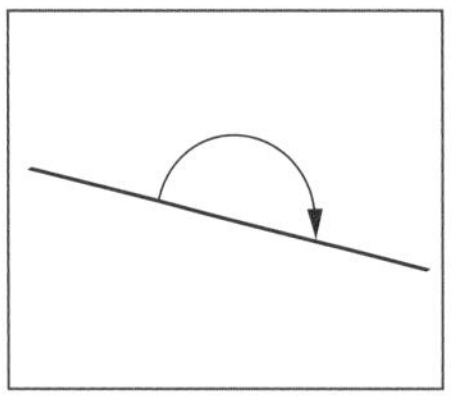

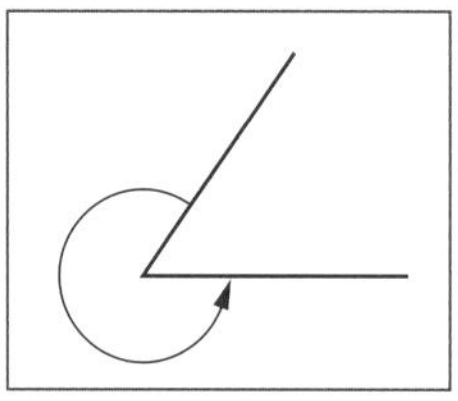

a I have 1 obtuse angle, 1 acute angle and 2 right angles. _____

A

b I have 3 acute angles. _____

B

c I have 3 right angles and 2 obtuse angles. _____

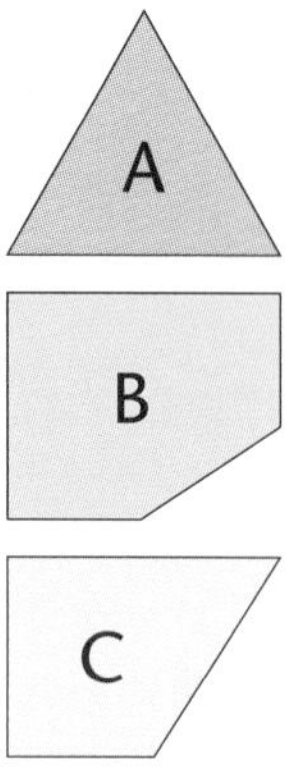

C

11 Draw a line to a label to match each angle created by the objects.

Acute angle

Right angle

Obtuse angle

2.5

S3.1 Students describe the defining geometric properties of families of 3D shapes, model 3D shapes using nets and other representations, and identify and describe the properties of specific families and subgroups of 2D shapes.

The base unit for measuring mass is the **kilogram**. The symbol for kilogram is **kg**. Everyday objects such as groceries are measured in kilograms. The prefix **kilo** means 1000. There are 1000 grams in a kilogram.

12 Hold a 1-kg mass, then use the hefting technique to identify objects that you estimate are less than, more than and about 1 kilogram.

A rockmelon is about 1 kilogram.

Less than 1 kg	About 1 kg	More than 1 kg

13 Use a pan balance to identify three objects to suit each category. You may like to check the items you hefted in the question above.

Less than 1 kg	About 1 kg	More than 1 kg

14 Use 5-kg kitchen scales to find the mass of the items to the nearest kilogram. Estimate the mass first by hefting.

	Item	Estimate	Mass
a	Phone book		
b	2 L of water		
c	Crash helmet		
d	Ream of A4 paper		
e	Sticky tape		
f	1 L of water		
g	10 books		

M3.1 Students identify and use equivalent forms of standard units when measuring, comparing and ordering, and estimate using a range of personal referents.

25

Split strategy for addition

1 Add the following numbers mentally by adding the tens part then the ones part. The first one is done for you.

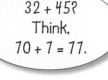

32 + 45?
Think,
70 + 7 = 77.

a 35 + 27 becomes | 50 | + | 12 | = | 62 |

b 35 + 46 becomes | | + | | = | |

c 37 + 36 becomes | | + | | = | |

d 56 + 27 becomes | | + | | = | |

e 87 + 96 becomes | | + | | = | |

f 126 + 38 becomes | 150 | + | 14 | = | |

g 147 + 35 becomes | | + | | = | |

h 215 + 69 becomes | | + | | = | |

i 347 + 25 becomes | | + | | = | |

2 Add the following numbers by adding the hundreds, then the tens and then the ones. The first one is done for you.

Add the hundreds, tens and ones separately.
415 + 234 = ?
Think 400 + 200 plus 10 + 30 plus 5 + 4 equals 649.

a 235 + 243 becomes | 400 | + | 70 | + | 8 | = | 478 |

b 423 + 264 becomes | | + | | + | | = | |

c 327 + 352 becomes | | + | | + | | = | |

d 316 + 273 becomes | | + | | + | | = | |

e 546 + 321 becomes | | + | | + | | = | |

f 624 + 255 becomes | | + | | + | | = | |

g 354 + 132 becomes | | + | | + | | = | |

3 Write a story problem to suit this number sentence.

365 + 70 =

_____

_____

_____

N3.2 Students identify and solve addition and subtraction problems involving whole numbers, and decimal fractions in context, selecting from a range of computation methods, strategies and known number facts.

Exploring fractions

Fractions can look different but still have the same value.

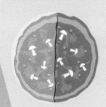

Jan ate $\frac{1}{2}$ of a pizza.

Susan ate $\frac{4}{8}$ of a pizza.

They ate the same amount.

4 Shade and record a fraction that looks different but still has the same value as the one given.

a  $\frac{1}{2}$ 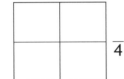 $\frac{}{4}$

d $\frac{1}{2}$ 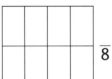 $\frac{}{8}$

b 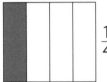 $\frac{1}{4}$ 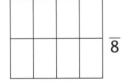 $\frac{}{8}$

e $\frac{3}{4}$ 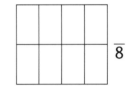 $\frac{}{8}$

c  $\frac{1}{2}$ 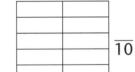 $\frac{}{10}$

f 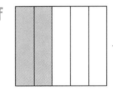 $\frac{2}{5}$ $\frac{}{10}$

5 Jordan cut some strips of paper then folded and labelled them to make fractions.

	$\frac{1}{2}$		

$\frac{1}{4}$	$\frac{2}{4}$	$\frac{3}{4}$	

$\frac{1}{8}$	$\frac{2}{8}$	$\frac{3}{8}$	$\frac{4}{8}$	$\frac{5}{8}$	$\frac{6}{8}$	$\frac{7}{8}$	

$\frac{1}{5}$	$\frac{2}{5}$	$\frac{3}{5}$	$\frac{4}{5}$	

$\frac{1}{10}$	$\frac{2}{10}$	$\frac{3}{10}$	$\frac{4}{10}$	$\frac{5}{10}$	$\frac{6}{10}$	$\frac{7}{10}$	$\frac{8}{10}$	$\frac{9}{10}$	

Study the strips of paper to decide whether the fractions below are equal to each other, greater than or less than the other. Use the >, = or < symbols to compare each pair of fractions.

$\frac{3}{5} > \frac{1}{2}$

a $\frac{2}{8}$ ☐ $\frac{1}{2}$ c $\frac{7}{10}$ ☐ $\frac{1}{2}$ e $\frac{1}{2}$ ☐ $\frac{2}{5}$ g $\frac{3}{4}$ ☐ $\frac{6}{8}$

b $\frac{1}{2}$ ☐ $\frac{5}{10}$ d $\frac{3}{4}$ ☐ $\frac{6}{8}$ f $\frac{1}{5}$ ☐ $\frac{3}{10}$ h $\frac{4}{8}$ ☐ $\frac{1}{2}$

N3.1 Students compare, order and represent whole numbers to 9 999 and common and decimal fractions, calculate cash transactions and describe other methods of payment.

27

Kim thought that Holdens were the most popular cars. While at the shops she observed the 7 most popular cars and made a picture graph to represent her observations.

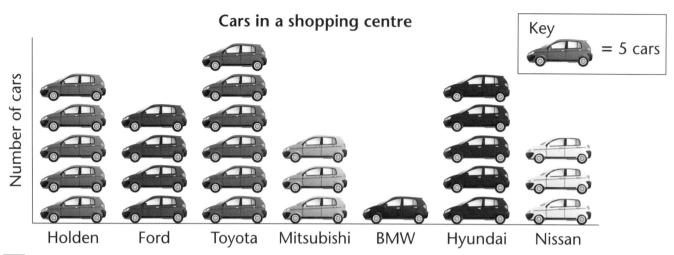

Cars in a shopping centre

Key
= 5 cars

Number of cars

Holden   Ford   Toyota   Mitsubishi   BMW   Hyundai   Nissan

6 Use the information in the key and the picture graph to answer these questions.

a How many cars in the car park? _____

b Which brand was seen the most? _____

c What brand was observed the least? _____

d How many Holdens were in the car park? _____

e How accurate was Kim's prediction? _____

7 Observe the types of cars passing your school. Record the number of each in the table below.

Sedans	Sports	4WD	Utilities	Vans	Trucks

8 Construct a picture graph to represent your data. Use a key so that each picture represents more than one car.

= [ ] vehicles

Cars passing the school

Number of vehicles

Sedans   Sports   4WD   Utilities   Vans   Trucks

CD3.2 Students design and trial a variety of data collection methods and use existing sources of data to investigate their own and others' questions, organise data and create suitable displays identifying and interpreting elements of the displays.

Square centimetres

9 Use the one-square-centimetre grid paper to design shapes of the given areas.

a 4 square centimetres	b 6 square centimetres	c 12 square centimetres

10 How many shapes can you make with an area of 9 square centimetres? Sketch them below.

11 Jessica has placed some centicubes on these shapes to help estimate their areas. Estimate and then measure the area of each shape.

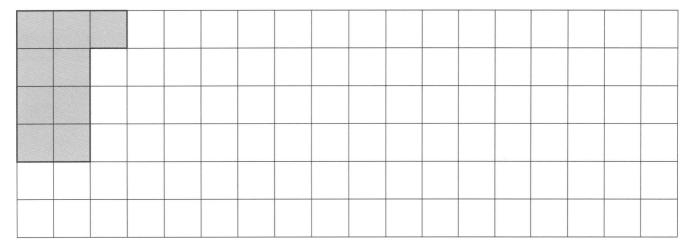

a

Est.	
Area	

b

Est.	
Area	

c

Est.	
Area	

M3.1 Students identify and use equivalent forms of standard units when measuring, comparing and ordering, and estimate using a range of personal referents.

29

Revising 3-digit subtraction/regrouping

- 5 ones take away
  6 ones can't be done
  at the moment so
  trade a ten for
  10 ones.
  15 take away 6.
- Trade 100 for 10 tens.
  14 tens take away 8 tens.
- 6 hundreds take away 0 hundreds.

Hund	Tens	Ones
6	14	15
7̶	5̶	5
−	8	6
6	6	9

1 Solve the algorithms.

a

Hund	Tens	Ones
8	4	5
−	9	6

b

Hund	Tens	Ones
7	5	3
−	7	4

c

Hund	Tens	Ones
6	3	5
−	5	7

d

Hund	Tens	Ones
9	5	2
−	9	3

e

Hund	Tens	Ones
1	3	3
−	5	5

f

Hund	Tens	Ones
5	2	1
−	7	3

g

Hund	Tens	Ones
3	4	1
−	5	2

h

Hund	Tens	Ones
6	2	3
−	4	5

i

Hund	Tens	Ones
6	3	3
− 5	8	2

j

Hund	Tens	Ones
3	2	2
− 2	7	5

k

Hund	Tens	Ones
5	1	8
− 4	8	4

l

Hund	Tens	Ones
7	5	9
− 6	7	9

m

Hund	Tens	Ones
8	2	0
− 7	1	8

2 Use a calculator to check your answers to Question 1.

3 Use any strategy you wish to solve the problems below.

Canberra to Sydney	288 km	Sydney to Melbourne	877 km
Canberra to Melbourne	647 km	Sydney to Brisbane	960 km

Brisbane to Rockhampton 637 km

	Event	Strategy
a	Janice and William both left Sydney at the same time. How much further did Janice travel on her trip to Melbourne compared to William who went to Canberra?	
b	How many more kilometres is the trip from Sydney to Melbourne compared with Brisbane to Rockhampton?	
c	How much further does a person living in Melbourne have to travel on a trip to Canberra compared to a person living in Sydney?	
d	How far is it from Melbourne to Brisbane via Canberra and Sydney?	

N3.2 Students identify and solve addition and subtraction problems involving whole numbers, and decimal fractions in context, selecting from a range of computation methods, strategies and known number facts.

Terms in number patterns

4 Continue the patterns that are modelled by the dice until the seventh term. Then record what you think would be the tenth term in the patterns.

a

  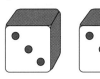

3	6	9	12				Tenth term

b

4	8	12	16				Tenth term

c

5	10	15	20				Tenth term

d

6	12	18	24				Tenth term

5 Creating square and triangular numbers.

a

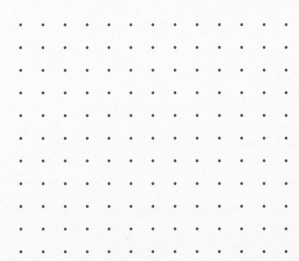

Create a square number larger than 16 that is also an even number.

b

Create a triangular number larger than 10 that is also an odd number.

PA3.1 Students create and continue number patterns, identify, describe and represent relationships between two quantities and use backtracking to reverse any one of the four operations.

6 Colour each object and matching description the same colour.

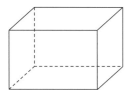

> **a** I am a prism that has 6 rectangular faces.

> **b** I am a pyramid that the Ancient Egyptians built. I have a square base.

> **c** I am a prism with 6 square faces.

> **d** I have a circular base and one vertex.

> **e** I am an object with 2 circles as bases.

> **f** I am a perfectly round 3D object.

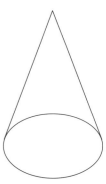
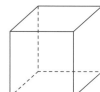

7 Model some of these objects from materials like matchsticks, toothpicks, plasticine and playdough.

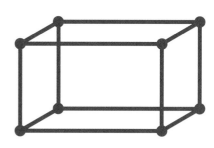

8 Describe the difference between a prism and a pyramid.

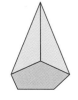
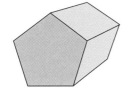

_____

_____

_____

S3.1 Students describe the defining geometric properties of families of 3D shapes, model 3D shapes using nets and other representations, and identify and describe the properties of specific families and subgroups of 2D shapes.

Volume can be measured in **litres**. The symbol for litres is **L**.

Find a 1-litre container like the ones shown below for use on this page of work.

9 Estimate first, then tally the number of times the egg cup, glass, cup and mug have to be filled in order to fill one of your 1-litre containers.

Estimate				
Tally				
Total				

10 Estimate, then measure how many litres of water each container holds (other containers can be substituted).

Container	Estimate	Litres

Container	Estimate	Litres

11 Discuss with a friend why you think there is a need for a unit smaller than a litre. Listen to their reasons to see if they are the same as yours.

M3.1 Students identify and use equivalent forms of standard units when measuring, comparing and ordering, and estimate using a range of personal referents.

33

Empty number lines

1 Karim had to solve this problem using a number line. This is what he did.

> Mr Lim had 139 sheep in one paddock and 33 in another. How many sheep did he have altogether?

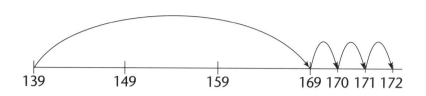

139   149   159   169 170 171 172

Solve these problems by filling in the number lines with the correct numbers, drawing the arcs on them to display the answers. The first two have been started for you.

a Mr Ross bought a jacket for $138 and a tie for $22. How much did he spend in total?

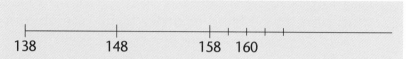

138   148   158 160

b Joanne saved $254 in March and $33 in April. How much did she save in the two-month period?

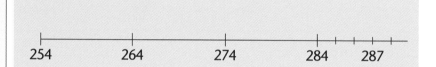

254   264   274   284   287

c Van Tran had $181 in the bank but spent $85 on new clothes. How much did she have left?

181

d Jaques had a collection of 349 football cards and bought 42 more. How many cards does he now have?

349

e Meagan had 166 raffle tickets but sold 34 of them on the weekend. How many does she have left?

f Tom made 116 cakes to sell at the fete but it rained and he only sold 23. How many did he have left?

g Harry had 157 shells that he found on the beach but he threw 33 away. How many does he have left?

h Jack saved $124 and Jill saved $32. How much money do they have altogether?

N3.2 Students identify and solve addition and subtraction problems involving whole numbers, and decimal fractions in context, selecting from a range of computation methods, strategies and known number facts.   **PA 3.2**

Expanding and ordering numbers

2 Write the numbers on the place value chart.

Number	Thousands	Hundreds	Tens	Ones
a 800		8	0	0
b 7 296				
c 2 307				
d 60				
e 5 207				
f 1 406				
g 6 237				

3 Place each set of numbers in descending order.

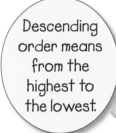

Descending order means from the highest to the lowest.

a 8 507, 7 503, 5 073, 3 057 _____

b 2 645, 3 658, 1 999, 2 500 _____

c 2 907, 8 436, 3 541, 2 657 _____

d 3 524, 5 234, 2 453, 4 532 _____

e 837, 238, 1 438, 2 745 _____

4 Write the largest number you can using the digits supplied.

a 3, 5, 6 _____   b 7, 0, 9 _____   c 3, 4, 2 _____   d 1, 3, 6, 5 _____   e 2, 7, 3, 4 _____

5 Expand each number. The first one has been done for you.

4 677 = 4 000 + 600 + 70 + 7

a 4 527 = [4 000] + [500] + [20] + [7]

b 5 436 = [ ] + [ ] + [ ] + [ ]

c 6 748 = [ ] + [ ] + [ ] + [ ]

d 6 740 = [ ] + [ ] + [ ] + [ ]

e 8 407 = [ ] + [ ] + [ ] + [ ]

f 7 987 = [ ] + [ ] + [ ] + [ ]

g 8 579 = [ ] + [ ] + [ ] + [ ]

N3.1 Students compare, order and represent whole numbers to 9 999 and common and decimal fractions, calculate cash transactions and describe other methods of payment.

35

North, south, east and west

6 Use the map to solve the direction questions below.

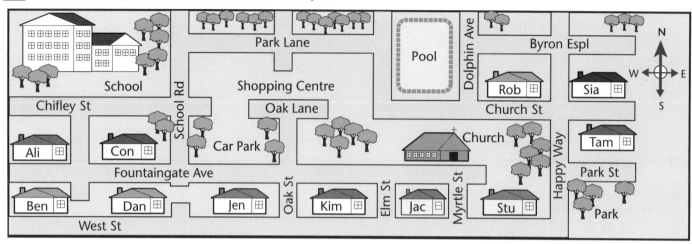

Which direction is it from:

a Stu's house to Rob's house? _____

b Jen's house to Kim's house? _____

c Jac's house to the church? _____

d Sia's house to Tam's house? _____

e Ali's house to Ben's house? _____

f The park to Sia's house? _____

g The pool to Jac's house? _____

h The school to the pool? _____

i Ali's house to the church? _____

j Jen's house to the car park? _____

k Write a set of directions explaining how to get from the park to the pool.

_____

_____

_____

7 Follow the directions from X to find the secret letter.

a Go north 4 spaces.

b Go east 4 spaces.

c Go south 3 spaces.

d Go east 3 spaces.

e Go north 4 spaces.

f Go west 2 spaces.

g Go south 3 spaces.

h What was the letter? _____

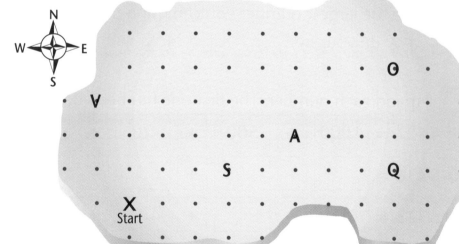

8 Give a set of directions explaining how to get to S from O.

_____

_____

_____

S3.2 Students interpret and create maps and plans using a range of conventions, describe locations and give directions using major compass points, angles and grids.

9 Jo said that if a dice is rolled 40 times every number from 1 to 6 will come up the same number of times.

a Do you agree with her? _____

b Predict the number you think will come up the most. _____

10 Roll a dice 40 times and tally the number of times each dice face lands, then construct a bar graph.

Dice	Tally 𝍷
⚀	
⚁	
⚂	
⚃	
⚄	
⚅	

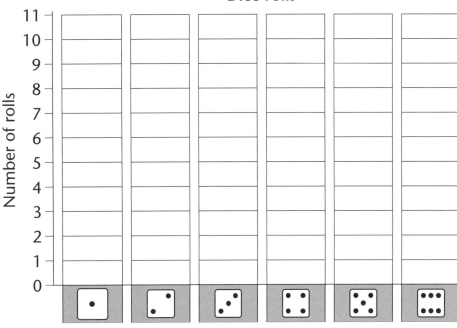

Dice rolls

11 Which face came up the most? _____

12 Which face came up the least? _____

13 Do your results match your friend's? _____

14 Why do you think they may be different? _____

15 Use these words to describe the chance that each event has of happening.

> possibly, probably, likely, unlikely, maybe, might, never, always, fifty-fifty, even chance, pretty sure, certain

	Event	Chance
a	I'll turn 11 next year.	
b	My sister becomes School Captain.	
c	The next traffic light we come to will be red.	
d	Our teacher will be away tomorrow.	
e	My pencil breaks today.	
f	I get 100% in my spelling test.	

CD3.1 Students identify all possible outcomes of familiar situations or actions and, for these sample spaces, order the likelihood of occurrence of the identified outcomes using experimental data.

37

Diagnostic review 1

PART 1

a Expand the numbers by writing them in the place-value grid.

Number	Thou	Hund	Tens	Ones
4 326				
5 279				
6 380				
4 206				
1 702				

b Write the number two thousand three hundred and twenty-six. ☐

c Order the numbers from smallest to largest.

1 357	7 537	3 571			

PART 2

Complete these additions and subtractions.

a 50 + 40 = ☐

b 80 − 50 = ☐

c 230 + 60 = ☐

d 150 − 80 = ☐

Use the split strategy to add these.

e 423 + 254 ☐ + ☐ + ☐ = ☐

f 537 + 218 ☐ + ☐ + ☐ = ☐

g 645 + 237 ☐ + ☐ + ☐ = ☐

h 328 + 479 ☐ + ☐ + ☐ = ☐

PART 3

Complete the grid.

×	2	3	4	5	6	7
a 2						
b 4						
c 7						
d 5						

Multiply these numbers

e $2 \times 40 =$ _____ **f** $2 \times 50 =$ _____

g $4 \times 40 =$ _____ **h** $4 \times 30 =$ _____

i $5 \times 40 =$ _____ **j** $7 \times 40 =$ _____

Complete these turnarounds.

k $50 \times 3 =$ ☐ , $3 \times$ ☐ $= 150$

l $60 \times 4 =$ ☐ , $4 \times$ ☐ $= 240$

m $42 \times 5 =$ ☐ , ☐ $\times 42 = 210$

Complete these number sentences.

n 2 × 8 × 5 = ☐

o 5 × 7 × 4 = ☐

PART 4

Complete the missing labels on the number lines.

a 0 ☐ $\frac{2}{4}$ ☐ 1

b 0 ☐ $\frac{2}{5}$ ☐ ☐ 1

Shade the larger fraction.

c $\frac{1}{5}$ or $\frac{1}{4}$ **d** $\frac{3}{4}$ or $\frac{3}{5}$

Say why your choice is larger in part **d**.

_____

Diagnostic review 1

PART 5

Draw an example of each type of angle.

a	Right angle	
b	Acute angle	
c	Obtuse angle	

PART 6

S3.1

Draw a line to match each shape with its name.

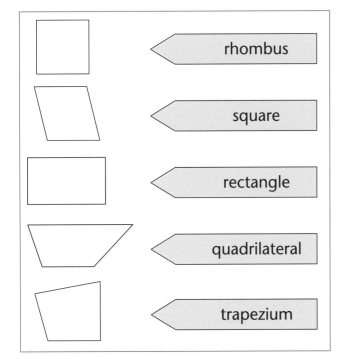

rhombus

square

rectangle

quadrilateral

trapezium

PART 7

S3.2

a What direction is it from Sam's house to Eve's house?

_____

b What direction is it from Eve's house to Lee's house?

_____

PART 8

M3.1

Measure the length of the pencil.

a  ☐ cm

Measure the perimeters in centimetres.

b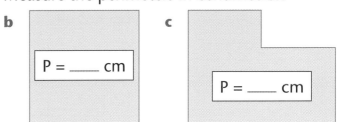

P = _____ cm

c

P = _____ cm

PART 9

M3.1

Record the area of each shape in square centimetres.

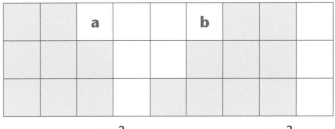

a   b

_____ cm² _____ cm²

PART 10

M3.1

Draw a line to match each object to a label.

| Less than 1 kg | About 1 kg | More than 1 kg |

PART 11

CD3.1

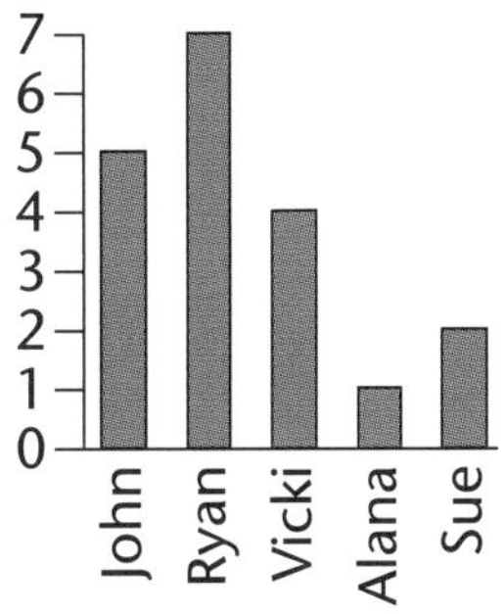

a Who had the most marbles? _____

b Who had 4 marbles? _____

c Who had 2 marbles? _____

d Who had 4 more marbles than Alana? _____

Multiplication strategies

Known multiplication facts can be used to find unknown facts.
E.g. 12 × 6? Think! 10 × 6 = 60 + 6 + 6 = 72.

12 × 8 = ?
Think! 10 × 8
plus 8 + 8
equals 96.

1 Answer the multiplications.

a 11 × 6 = ☐   f 12 × 5 = ☐   k 13 × 5 = ☐

b 11 × 5 = ☐   g 12 × 6 = ☐   l 13 × 6 = ☐

c 11 × 7 = ☐   h 12 × 7 = ☐   m 14 × 4 = ☐

d 11 × 8 = ☐   i 12 × 8 = ☐   n 14 × 5 = ☐

e 12 × 4 = ☐   j 13 × 4 = ☐   o 14 × 6 = ☐

A strategy to use when multiplying by 4 is to double then double again.
E.g. 8 × 4? Think! Double 8 = 16 then double 16 = 32.

23 × 4 = ?
Double 23 = 46.
Double 46 = 92.

2 Use the 'double and double again' strategy to multiply by 4.

a 6 × 4 = ☐   e 14 × 4 = ☐   i 20 × 4 = ☐

b 8 × 4 = ☐   f 13 × 4 = ☐   j 22 × 4 = ☐

c 10 × 4 = ☐   g 15 × 4 = ☐   k 25 × 4 = ☐

d 12 × 4 = ☐   h 16 × 4 = ☐   l 30 × 4 = ☐

A strategy for multiplying by 6 is to multiply by 3 then double.
A strategy for multiplying by 5 is to multiply by 10 then halve.

16 × 5 = ?
Think!
16 × 10 = 160,
half of 160 = 80.

3 Use these strategies to answer the questions.

a 8 × 5 = ☐   f 15 × 5 = ☐   k 7 × 6 = ☐

b 9 × 5 = ☐   g 20 × 5 = ☐   l 9 × 6 = ☐

c 10 × 5 = ☐   h 4 × 6 = ☐   m 8 × 6 = ☐

d 12 × 5 = ☐   i 6 × 6 = ☐   n 10 × 6 = ☐

e 14 × 5 = ☐   j 5 × 6 = ☐   o 12 × 6 = ☐

N3.3 Students identify and solve multiplication and division problems involving whole numbers, and decimal fractions in context, selecting from a range of computation methods, strategies and known number facts.

Mixed numbers

A **mixed number** is a number that consists of a whole number and a fraction.

E.g. The model displays 1 whole and 1 half.

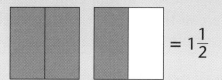 $= 1\frac{1}{2}$

4 Write the mixed numerals displayed by the shaded shapes below.

a 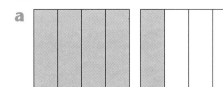 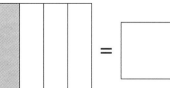 $=$

b 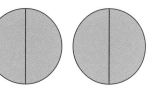 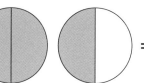 $=$

c 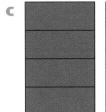 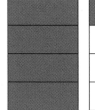 $=$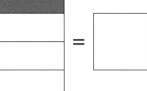

d   $=$

e 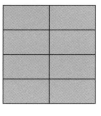 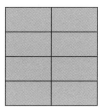 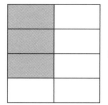 $=$

5 Shade the shapes to display the mixed numbers.

a $1\frac{3}{4}$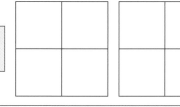

d $1\frac{4}{5}$

b $1\frac{3}{5}$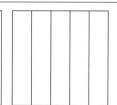

e $2\frac{3}{8}$

c $2\frac{7}{8}$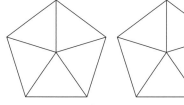

f $1\frac{3}{10}$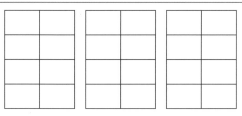

N3.1 Students compare, order and represent whole numbers to 9 999 and common and decimal fractions, calculate cash transactions and describe other methods of payment.

41

When a set of shapes fit together in a repeating pattern, without any gaps, we call the pattern a **tessellation**. Bathroom tiles and bricks are good examples of tessellations.

6 Classify the sets of shapes into tessellating and non-tessellating. Write your answers in the grid.

Tessellating				
Non-tessellating				

a

b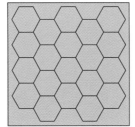

c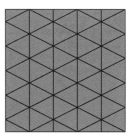

d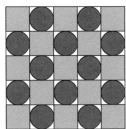

e

f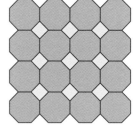

7 Continue the tessellations.

a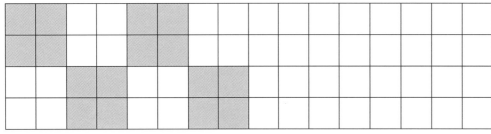

b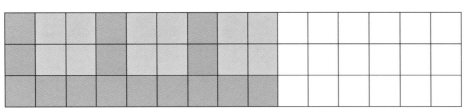

8 Create a tessellating pattern of your own and sketch it in the space provided.

S3.1 Students describe the defining geometric properties of families of 3D shapes, model 3D shapes using nets and other representations, and identify and describe the properties of specific families and subgroups of 2D shapes.

Remember! There are 1 000 millilitres in a litre. The symbol for millilitres is **mL**.

9 Order the following containers from smallest to largest by numbering them from 1 to 7 according to their capacity.

a · 375 mL
b · 300 ml
c · 200 mL
d · 1.25 L
e
f · 1 LITRE
g · 500 mL

10 Test to see how many times your coffee cup needs to be filled in order to fill a 1-litre milk carton. Colour the number of cups you used.

11 Use a medicine glass to find the volume of the containers.

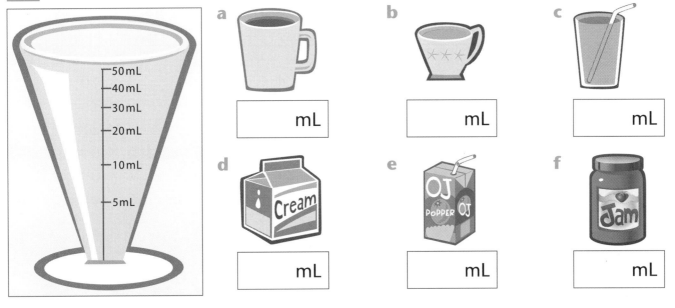

50 mL
40 mL
30 mL
20 mL
10 mL
5 mL

a ____ mL
b ____ mL
c ____ mL
d ____ mL
e ____ mL
f ____ mL

M3.1 Students identify and use equivalent forms of standard units when measuring, comparing and ordering, and estimate using a range of personal referents.

43

Division strategies

1 Use halving skills to divide by 2.

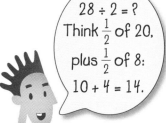

a 12 ÷ 2 = ☐   e 30 ÷ 2 = ☐   i 48 ÷ 2 = ☐

b 16 ÷ 2 = ☐   f 36 ÷ 2 = ☐   j 38 ÷ 2 = ☐

c 20 ÷ 2 = ☐   g 44 ÷ 2 = ☐   k 46 ÷ 2 = ☐

d 24 ÷ 2 = ☐   h 66 ÷ 2 = ☐   l 86 ÷ 2 = ☐

> 28 ÷ 2 = ?
> Think $\frac{1}{2}$ of 20, plus $\frac{1}{2}$ of 8:
> 10 + 4 = 14.

2 Use the 'halve and halve again' strategy to divide by 4.

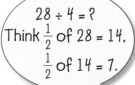

a 12 ÷ 4 = ☐   e 32 ÷ 4 = ☐   i 44 ÷ 4 = ☐

b 20 ÷ 4 = ☐   f 40 ÷ 4 = ☐   j 8 ÷ 4 = ☐

c 16 ÷ 4 = ☐   g 36 ÷ 4 = ☐   k 24 ÷ 4 = ☐

d 28 ÷ 4 = ☐   h 48 ÷ 4 = ☐   l 52 ÷ 4 = ☐

> 28 ÷ 4 = ?
> Think $\frac{1}{2}$ of 28 = 14.
> $\frac{1}{2}$ of 14 = 7.

3 Solve these problems using any strategy you wish.

a	Samantha is buying a skateboard for $160. To pay for it she has to make 8 equal payments. How much will she pay each time?		
b	If the trip to the coast is 864 km, how far would we have travelled if we have completed $\frac{1}{4}$ of the trip?	 The Coast 864 km	

4 Complete the number cross.

Across

2 4 × 6 =
4 24 ÷ 2 =
5 40 ÷ 4 =
7 15 ÷ 5 =
8 21 ÷ 3 =
10 25 ÷ 5 =
11 50 ÷ 5 =
13 28 ÷ 4 =
14 24 ÷ 6 =
16 36 ÷ 3 =
17 7 × 4 =

Down

1 3 × 4 =
3 3 × 10 =
6 9 × 3 =
9 7 × 3 =
12 6 × 6 =
15 7 × 6 =

> Did you know that
> 3 × 4 = 24 ÷ 2?

N3.3 Students identify and solve multiplication and division problems involving whole numbers, and decimal fractions in context, selecting from a range of computation methods, strategies and known number facts.

5 Use the array to answer these 8's facts.

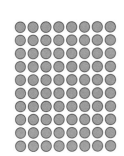

a 1 × 8 = ☐   e 5 × 8 = ☐   i 9 × 8 = ☐

b 2 × 8 = ☐   f 6 × 8 = ☐   j 10 × 8 = ☐

c 3 × 8 = ☐   g 7 × 8 = ☐

d 4 × 8 = ☐   h 8 × 8 = ☐

6 Use the 'double, double, then double again' strategy to solve these 8's facts.

a 3 × 8 = ☐   f 9 × 8 = ☐

b 4 × 8 = ☐   g 11 × 8 = ☐

c 7 × 8 = ☐   h 15 × 8 = ☐

d 5 × 8 = ☐   i 13 × 8 = ☐

e 8 × 8 = ☐   j 14 × 8 = ☐

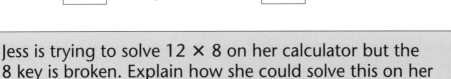

6 × 8 = ?
Think, double 6 = 12,
double 12 = 24,
double 24 = 48.

7 Jess is trying to solve 12 × 8 on her calculator but the 8 key is broken. Explain how she could solve this on her calculator by multiplying by 4 or 2 instead of 8.

8 Crack the secret code by exchanging answers for letters.

A	B	C	D	E	F	G	I	J	K	Q	R	S	T	U	V
28	72	21	24	40	63	42	80	32	36	35	56	64	49	50	10

1 6 × 7 = ☐   5 7 × 7 = ☐   9 60 − 4 = ☐   13 50 + 6 = ☐

2 7 × 8 = ☐   6 9 × 8 = ☐   10 10 × 8 = ☐   14 4 × 10 = ☐

3 5 × 8 = ☐   7 7 × 4 = ☐   11 8 × 5 = ☐   15 10 × 4 = ☐

4 4 × 7 = ☐   8 8 × 7 = ☐   12 53 + 3 = ☐   16 9 × 7 = ☐

1	2	3	4	5	6	7	8	9	10	11	12	13	14	15	16

N3.3 Students identify and solve multiplication and division problems involving whole numbers, and decimal fractions in context, selecting from a range of computation methods, strategies and known number facts.

45

Greece || TTTT | |
Italy || TTTT
Vietnam || TTTT | | |
Australia || TTTT | | | |
Lebanon || | | | |

9 Ms Evans did a survey in her class of the countries that her students' fathers come from. She asked her class to present the country of origin data in three different ways. Make a table from the tallied data Ms Evans supplied.

Greece	Italy	Vietnam	Australia	Lebanon
6				

10 Make a bar graph from the table you have made above.

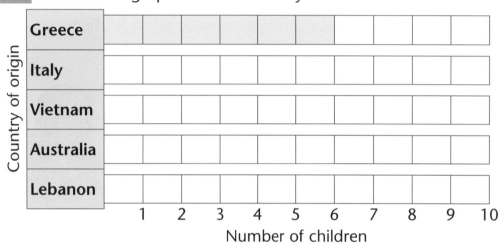

Country of origin

Greece										
Italy										
Vietnam										
Australia										
Lebanon										
	1	2	3	4	5	6	7	8	9	10

Number of children

11 In groups or pairs discuss how this data could be varied, such as by asking where mothers, grandfathers or neighbours came from.

Carry out your own survey and present your information in a table and as a graph.

Country					
Number					

Country of origin

	1	2	3	4	5	6	7	8	9	10

Number of children

My mum was born in _____

12 Was your data similar to the data collected by Ms Evans? _____

_____

CD3.2 Students design and trial a variety of data collection methods and use existing sources of data to investigate their own and others' questions, organise data and create suitable displays identifying and interpreting elements of the displays.

Time units

13 Complete and learn these facts.

a | 60 seconds = [ ]
b | 60 minutes = [ ]
c | 24 hours = [ ]
d | 7 days = [ ]
e | 1 fortnight = [ ]

f | 52 weeks = [ ]
g | 12 months = [ ]
h | 365 days = [ ]
i | 366 days = [ ]

14 Would you use seconds, minutes, hours, days, weeks, months or years to measure these periods of time?

a Lunch time [ ]

b Football match [ ]

c TV advertisement [ ]

d Christmas holidays [ ]

e Walk 10 metres [ ]

f Cricket match [ ]

g Your favourite song [ ]

h Spring [ ]

i Recess [ ]

j Your life [ ]

15 Use the greater than or less than symbols to make these sentences true.

> is the greater than symbol.
< is the less than symbol.

a 5 days is [ ] one week

b 75 minutes is [ ] 1 hour

c 12 hours is [ ] 1 day

d 59 seconds is [ ] 1 minute

e 25 hours is [ ] 1 day

f 90 seconds is [ ] 1 hour

g 13 days is [ ] 2 weeks

h 130 minutes is [ ] 2 hours

i 17 days is [ ] 2 weeks

j 36 hours is [ ] 2 days

16 Place these swimming times in order for the judges.

1 min 18 sec	1 min 19 sec	1 min 20 sec	1 min 12 sec	1 min 56 sec
		fourth		

M3.2 Students read, record and calculate with 12-hour time, and interpret calendars and simple timetables related to daily activities.

47

Multiplication—the 9's facts

1 Use the arrays to answer the 9's facts.

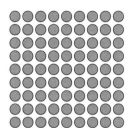

a 1 × 9 = ☐   e 5 × 9 = ☐   h 8 × 9 = ☐

b 2 × 9 = ☐   f 6 × 9 = ☐   i 9 × 9 = ☐

c 3 × 9 = ☐   g 7 × 9 = ☐   j 10 × 9 = ☐

d 4 × 9 = ☐

2 Circle the answers to the 9's facts on the chart.

3 Why would it be easy to spot a mistake in the 9's facts on the chart?
_____

4 Shade the answers to the 3's facts on the chart and continue the pattern until it makes to 90.

5 Can you see a relationship between the 3's facts and 9's facts? _____

6 What is it? _____

1	2	3	4	5	6	7	8	9	10
11	12	13	14	15	16	17	18	19	20
21	22	23	24	25	26	27	28	29	30
31	32	33	34	35	36	37	38	39	40
41	42	43	44	45	46	47	48	49	50
51	52	53	54	55	56	57	58	59	60
61	62	63	64	65	66	67	68	69	70
71	72	73	74	75	76	77	78	79	80
81	82	83	84	85	86	87	88	89	90

7 Complete the multiplication facts grid.

×	2	4	5	1	3	7	6	9	10	8
9										
10										
6										
7										
8										

Make sure you learn your multiplication facts. x3 x8 x7 x6 x9

8 Find the missing digits in these multiplication facts.

a 4 × 9 = ☐   g ☐ × 9 = 27

b ☐ × 6 = 54   h ☐ × 7 = 28

c 7 × ☐ = 63   i ☐ × 6 = 48

d 5 × ☐ = 45   j 7 × ☐ = 49

e 9 × ☐ = 81   k ☐ × 8 = 64

f ☐ × 9 = 72   l 7 × ☐ = 56

N3.3 Students identify and solve multiplication and division problems involving whole numbers, and decimal fractions in context, selecting from a range of computation methods, strategies and known number facts.

Mixed numbers are a combination of whole numbers and fractions.
E.g. 0, $\frac{1}{2}$, 1, $1\frac{1}{2}$, 2, $2\frac{1}{2}$, 3

9 Complete the number line to count by halves, quarters and fifths.

a

b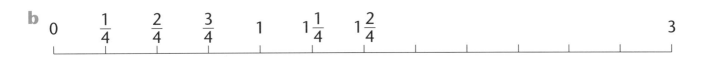

c

10 Draw a line to show where each fraction or mixed number belongs on the number line.

a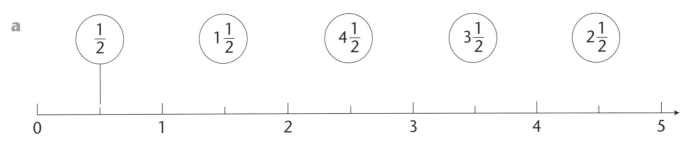

b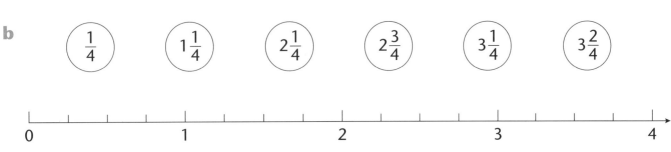

11 Write the missing mixed numbers on the T-shirts.

N3.1   PA3.1 Students create and continue number patterns, identify, describe and represent relationships
                              between two quantities and use backtracking to reverse any one of the four operations.

49

Prisms step 1	Prisms step 2	Pyramids step 1	Pyramids step 2
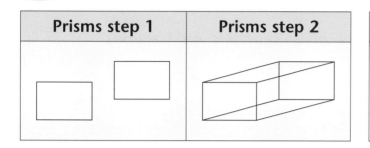			

12 Trace the bases of the objects first before joining their corners.

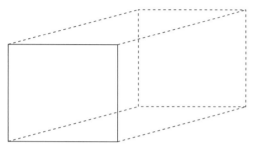 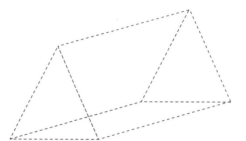 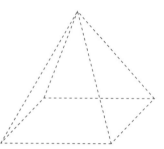

13 Draw these objects using dotted lines to show edges that are out of sight.

a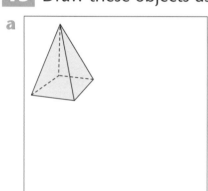

b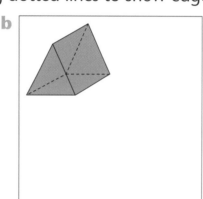

c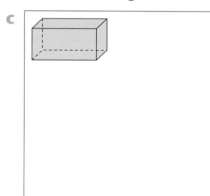

d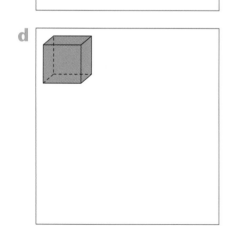

e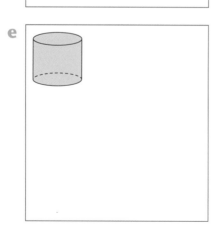

f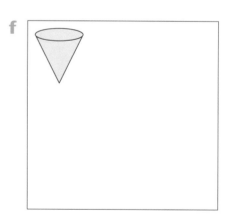

14 Is this shape a prism or a pyramid? _____

Why does the shape belong to that family? _____

_____

_____

S3.1 Students describe the defining geometric properties of families of 3D shapes, model 3D shapes using nets and other representations, and identify and describe the properties of specific families and subgroups of 2D shapes.

Grams are used to measure objects that are not very heavy.
The symbol for grams is **g**.

1 kilogram = 1000 grams

15 Collect three plastic jars with screw-on lids and make your own standard mass set.

a	Jar 1	Pour sand into Jar 1 until it balances a 100-g mass.	100 g
b	Jar 2	Pour sand into Jar 2 until it balances a 200-g mass.	200 g
c	Jar 3	Pour sand into Jar 3 until it balances a 500-g mass.	500 g

16 Use your standard mass set to find three objects which closely match each mass.

About 100 g	About 200 g	About 500 g

17 How many of each fruit or vegetable are needed to balance 1 kg?

	Item	Mass	Number
a	Tomato	100 g	
b	Apple	200 g	
c	Pineapple	500 g	
d	Potato	250 g	
e	Strawberry	50 g	

M3.1 Students identify and use equivalent forms of standard units when measuring, comparing and ordering, and estimate using a range of personal referents.

51

Compensation strategy/ rearranging numbers

The **compensation strategy** can be done by **rounding up** one of the numbers that you are adding. E.g. 354 + 38 = ? Think 354 + 40 = 394, then take off the 2 that was rounded up. The answer is 392.

1 Use this strategy to complete the additions. The first one has been done for you.

a 347 + 38 becomes | 347 | + | 40 | = | 387 | subtract | 2 | = | 385

b 455 + 19 becomes | | + | | = | | subtract | | = | _____

c 563 + 27 becomes | | + | | = | | subtract | | = | _____

d 245 + 29 becomes | | + | | = | | subtract | | = | _____

e 373 + 38 becomes | | + | | = | | subtract | | = | _____

f 166 + 39 becomes | | + | | = | | subtract | | = | _____

g 276 + 48 becomes | | + | | = | | subtract | | = | _____

When we are adding, part of one number can be used to round up the other. E.g. 463 + 28 becomes 461 + 30 = 491

2 Rearrange these numbers to make the additions easier.

a 235 + 49 becomes | | + | | = | |

b 364 + 57 becomes | | + | | = | |

c 543 + 687 becomes | | + | | = | |

d 596 + 257 becomes | | + | | = | |

e 646 + 299 becomes | | + | | = | |

238 + 47
Take 3 off 238
and add it to 47.
235 + 50 = 285

3 Use one of the mental strategies above to solve these additions.

a 637 + 29 = ____   c 339 + 51 = ____   e 356 + 233 = ____   g 333 + 348 = ____

b 258 + 38 = ____   d 247 + 58 = ____   f 277 + 422 = ____   h 572 + 209 = ____

4 Explain the strategy you would use to solve: 148 + 36 = [ ] .

N3.2 Students identify and solve addition and subtraction problems involving whole numbers, and decimal fractions in context, selecting from a range of computation methods, strategies and known number facts.

PA2.2

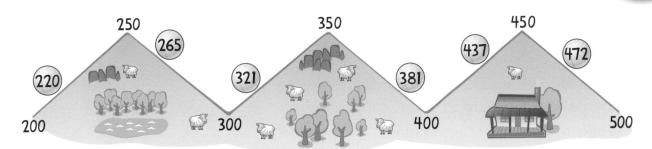

5 Which hundred would each ball roll to?

a (220) _____ b (321) _____ c (437) _____ d (472) _____ e (265) _____

6 Round off each number to the nearest 100. (Numbers ending in 50 are rounded up.)

a 203 _____ d 297 _____ g 387 _____ j 457 _____
b 449 _____ e 3259 _____ h 4567 _____ k 3529 _____
c 211 _____ f 3450 _____ i 7468 _____ l 9209 _____

7 Round each number to the nearest 100 in order to give an approximate solution.

a 498 + 103 ≈ _____ d 621 + 398 ≈ _____ g 698 + 203 ≈ _____

679 – 457?
Think
700 – 500
≈ 200.

b 967 − 454 ≈ _____ e 878 − 476 ≈ _____ h 679 − 457 ≈ _____

c 404 + 102 ≈ _____ f 375 + 211 ≈ _____ i 642 + 153 ≈ _____

8 Only consider the hundreds and tens to give an approximate answer to these subtractions. The first one has been done as an example.

	Question		Strategy	Approximate answer
a	786 − 259	becomes	78 tens − 25 tens = 53 tens	about 530
b	562 − 349	becomes		
c	683 − 478	becomes		
d	973 − 545	becomes		
e	858 − 247	becomes		
f	766 − 447	becomes		

Horizontal, vertical and oblique lines

Horizontal lines are parallel to the horizon.
The towel is in a horizontal position.
Vertical lines are at right angles to the horizon.
The parallel flags are in a vertical position.
Oblique lines are at an angle to the horizon.
The umbrella is in an oblique position.

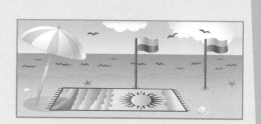

9 Label the lines as horizontal, vertical or oblique when compared to the horizon which has been drawn as a dotted line. Colour the space between any parallel lines.

a	b	c	d
e	f	g	h

10 Are these statements true or false?

a The top of the coffee table is a horizontal line. _____

b The sides of the picture are parallel to each other. _____

c The bottom of the picture frame is an oblique line. _____

d The angle of the fishing rod is oblique. _____

e The sides of the lounge are vertical lines. _____

f The books are all on oblique angles. _____

11 List some horizontal, vertical and oblique lines found in your classroom.

Horizontal	Vertical	Oblique

S3.1 Students describe the defining geometric properties of families of 3D shapes, model 3D shapes using nets and other representations, and identify and describe the properties of specific families and subgroups of 2D shapes.

More likely, less likely

James found a box of counters and laid them out in their colours.

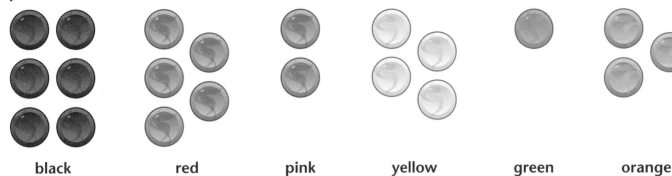

| black | red | pink | yellow | green | orange |

12 Write true or false for the statements based on the knowledge that James put all the counters back in the box and then selected one counter to show his mum.

True or False

a James is more likely to pick out a black counter than a red one. _____

b James is less likely to pick out a red one than an orange one. _____

c James is more likely to pick out a red one than a yellow one. _____

d James is more likely to pick out a pink one than a yellow one. _____

e James is less likely to pick out an orange one than a black one. _____

f James is more likely to pick out a yellow one than an orange one. _____

13 Order the colours from the one that is least likely to be taken out of the box to the one that is most likely.

_____

_____

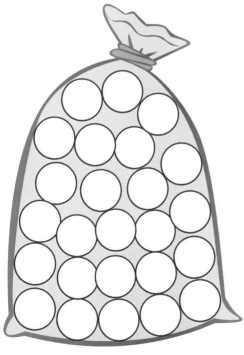

14 Place the marbles in the bag by writing a letter on each marble for each colour.

a There are 4 red marbles.

b There are 3 more green marbles than red ones.

c There are half as many blue marbles as red marbles.

d There are 2 more yellow marbles than green ones.

e The rest of the marbles are pink.

15 Answer the questions.

a Which is the most likely marble to be drawn out of the bag? _____

b Which is the least likely marble to be drawn out of the bag? _____

c Which is more likely to be drawn from the bag: a green or a red marble? _____

Addition to 9999

1 thousand plus 3 thousands plus 4 thousands equals 8 thousands.

5 ones plus 8 ones equals 13 ones. Exchange 10 ones for 1 ten then record the 3 in the ones column.

Thou	Hund	Tens	Ones
1	1	1	
3	8	5	5
+ 4	5	6	8
8	4	2	3

1 hundred plus 8 hundreds plus 5 hundreds equals 14 hundreds. Exchange 10 hundreds for 1 thousand then record the 4 in the hundreds column.

1 ten plus 5 tens plus 6 tens equals 12 tens. Exchange 10 tens for 1 hundred then record the 2 in the tens column.

1 Add the 4-digit numbers without trading.

a 3 6 3 3
+ 4 2 3 5

b 4 5 3 2
+ 3 3 4 7

c 1 2 3 4
+ 4 3 5 1

d 2 7 3 4
+ 7 2 6 5

e 3 4 5 6
+ 3 2 3 3

2 Add the 4-digit numbers with regrouping in the ones.

a 3 5 6 8
+ 4 2 1 4

b 3 2 1 9
+ 4 7 4 3

c 3 6 3 7
+ 4 0 4 6

d 5 6 5 8
+ 4 0 2 8

e 6 7 6 7
+   2 1 5

3 Add the 4-digit numbers with regrouping in the ones and tens.

a 3 5 6 2
+ 4 2 5 3

b 3 5 7 9
+   3 7 0

c 2 5 8 6
+ 5 3 0 7

d 3 6 7 8
+ 2 1 4 4

e 5 4 6 8
+ 3 2 5 5

4 Solve these problems.

a	When the train left the city there were 165 passengers on board. At the first station 86 people boarded the train and another 73 boarded at the next station. How many people altogether were on the train?		
b	A driver had 3 parcels to deliver. Their masses were 326 kg, 438 kg and 160 kg. What is the total mass of the parcels?		
c	A hired car was driven 1422 km from Sydney to Adelaide and then another 728 km to Melbourne. What is the total distance travelled?		
d	On Saturday 1378 people attended the fete and on Sunday 2 486 people attended. How many people attended the fete during the weekend?		

N3.2 Students identify and solve addition and subtraction problems involving whole numbers, and decimal fractions in context, selecting from a range of computation methods, strategies and known number facts.

Division facts

Another way to express division is the division symbol $\overline{}$.

$3\overline{)12}$ with 4 above, means $12 \div 3 = 4$

5 Complete the division facts.

a $3\overline{)9}$	d $4\overline{)24}$	g $5\overline{)30}$	j $6\overline{)36}$	m $7\overline{)35}$
b $4\overline{)16}$	e $3\overline{)12}$	h $6\overline{)18}$	k $7\overline{)42}$	n $8\overline{)80}$
c $5\overline{)35}$	f $4\overline{)28}$	i $5\overline{)40}$	l $8\overline{)56}$	o $9\overline{)81}$

6 Solve these division facts using known multiplication facts.

a $18 \div 3 = \boxed{}$ e $24 \div 6 = \boxed{}$ i $21 \div 7 = \boxed{}$

b $15 \div 3 = \boxed{}$ f $18 \div 2 = \boxed{}$ j $32 \div 8 = \boxed{}$

c $36 \div 6 = \boxed{}$ g $20 \div 4 = \boxed{}$ k $45 \div 9 = \boxed{}$

d $40 \div 5 = \boxed{}$ h $30 \div 6 = \boxed{}$ l $49 \div 7 = \boxed{}$

$12 \div 3 = ?$
Think
$4 \times 3 = 12.$

7 Think of a multiplication fact that will help complete each division fact.

$\boxed{} \times \boxed{} = \boxed{}$	$\boxed{} \times \boxed{} = \boxed{}$	$\boxed{} \times \boxed{} = \boxed{}$
$32 \div 4 = \boxed{}$	$35 \div 5 = \boxed{}$	$42 \div 6 = \boxed{}$
$32 \div 8 = \boxed{}$	$35 \div 7 = \boxed{}$	$22 \div 7 = \boxed{}$

ZAPPA $4 Spooky $6 TURKEY $12 the great aussie lolly WILD $8

8 Louise spent $48 on show bags. How many of each bag could she buy with $48?

a Zappa _____   b Spooky _____   c Wild _____   d Turkey _____

9 How many division number sentences can you write with an answer of 6?

N3.3 Students identify and solve multiplication and division problems involving whole numbers, and decimal fractions in context, selecting from a range of computation methods, strategies and known number facts.

57

Reading maps, using a legend

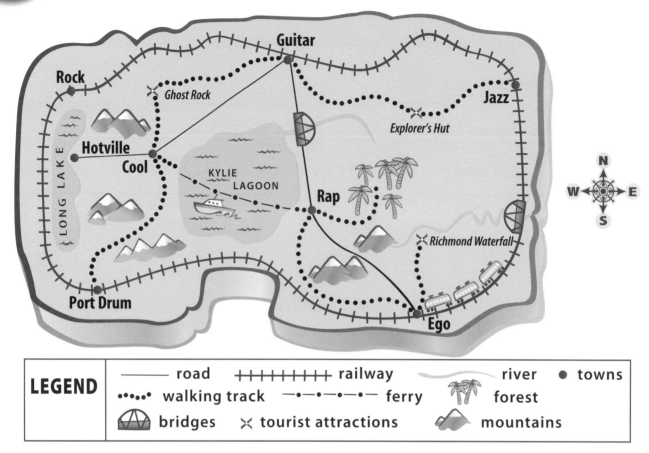

LEGEND
——— road ++++++++ railway ～～～ river • towns
•••••• walking track –•–•–•– ferry 🌴 forest
⌂ bridges ✕ tourist attractions ⛰ mountains

10 Use the legend to answer the questions.

a How many towns are on the island? _____

b How many tourist attractions are marked on the map? _____

c How many bridges are on the island? _____

d Is it possible to drive to the tourist attractions? _____

e Which town can only be reached by road? _____

f How many rivers are on the island? _____

g Which town can only be reached by train? _____

h Name one town that can be reached by road, train and by walking. _____

i If I was at Guitar, would I have to pass through Rap to get to Ego? _____

j Which town is south of Rock? _____

k Which two towns does the ferry help join? _____

Legends help people read maps.

11 Use north, south, east and west to give the directions from:

a Hotville to Cool _____
b Jazz to Rock _____
c Rock to Hotville _____
d Explorer's Hut to Ego _____
e Cool to Ghost Rock _____
f Kylie Lagoon to Long Lake _____

S3.2 Students interpret and create maps and plans using a range of conventions, describe locations and give directions using major compass points, angles and grids.

12 Estimate the area of the top surface of these objects. Place each object on top of 1-cm grid paper or use a clear plastic grid overlay to measure the area.

Object	Estimate	Square centimetres

13 Peter made a shape that covered an area of 16 square centimetres. He said that all shapes with an area of 16 square centimetres also have a perimeter of 16 cm. Janice said that he was wrong.

a Draw 2 more shapes with an area of 16 square centimetres.

b Who was correct? _____

Peter's shape

M3.1 Students identify and use equivalent forms of standard units when measuring, comparing and ordering, and estimate using a range of personal referents.

59

Multiplying mentally

One strategy used to multiply a 2-digit number by a 1-digit number is to multiply the **tens** then the **ones**.

150   30

E.g. 36 × 5 becomes 30 × 5 plus 6 × 5 = 180

1 Use this strategy to solve the questions. The first one is done for you.

	Question		Working		Answer
a	19 × 6	becomes	10 × 6 = 60 plus 9 × 6 = 54	=	114
b	18 × 5	becomes		=	
c	21 × 3	becomes		=	
d	33 × 5	becomes		=	
e	41 × 7	becomes		=	
f	39 × 5	becomes		=	
g	28 × 4	becomes		=	
h	52 × 8	becomes		=	
i	58 × 3	becomes		=	

Estimation is often used to check if the answers to multiplications are reasonable.

2 Check to see if your answers to Question 1 are reasonable by rounding the 2-digit number to the nearest 10 to find an approximate answer.

20

a 19 × 6 ≈ 120

b 18 × 5 ≈

c 21 × 3 ≈

d 33 × 5 ≈

e 41 × 7 ≈

f 39 × 5 ≈

g 28 × 4 ≈

h 52 × 8 ≈

i 58 × 3 ≈

3 Solve these problems.

	Problem	Working
a	How far did the relay team run if each of the 4 runners ran 19 kilometres?	
b	How much would it cost to buy 5 paintings at a cost of $48 each?	

N3.3 Students identify and solve multiplication and division problems involving whole numbers, and decimal fractions in context, selecting from a range of computation methods, strategies and known number facts.

Decimals and fractions

23 hundredths can be written as a fraction $\frac{23}{100}$ and as a decimal **0.23**.

4 Record each shaded hundredth grid as a fraction and a decimal.

a 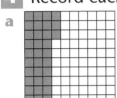 b 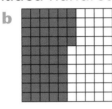 c 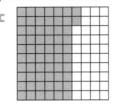 d 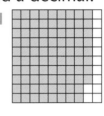 e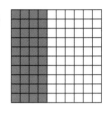

$\frac{}{100}$ [ ] . [ ] $\frac{}{100}$ [ ] . [ ] $\frac{}{100}$ [ ] . [ ] $\frac{}{100}$ [ ] . [ ] $\frac{}{100}$ [ ] . [ ]

5 Shade the materials to match the fractions or decimals.

a 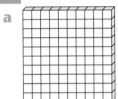 b 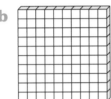 c 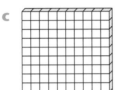 d 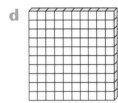 e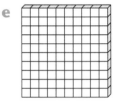

0.35 $\frac{38}{100}$ 0.28 $\frac{72}{100}$ 0.57

6 Shade the materials to match the decimal as closely as you can.

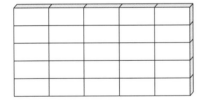      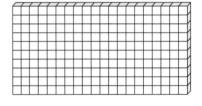

0.80   0.60   0.40

7 Draw a line to show where each decimal belongs on the number lines.

a   0.50   0.25   0.75   0.60   0.85

0 ———————————————————————————————— 1

b   1.20   1.50   0.80   1.90   1.30

0 ———————————————— 1 ———————————————— 2

N3.1 Students compare, order and represent whole numbers to 9 999 and common and decimal fractions, calculate cash transactions and describe other methods of payment.

61

UNIT 15 Geometric symbols

Symbols are used to label shapes and angles:
∥ strokes are used to show sides of equal length
> arrows are used to show parallel sides
□ a small square is used to show right angles.

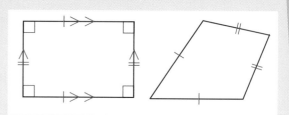

8 Measure the sides of the shapes below to identify sides of equal length.
A single stroke is used for the first set of equal sides. Double strokes are used if there
is another set of sides that are equal to each other.

a 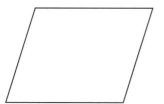 b 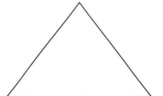 c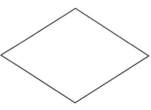

How many equal sides does a rhombus have?

9 Use arrows to identify parallel sides on these shapes..

a 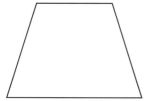 b 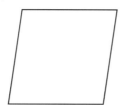 c

Do all quadrilaterals have parallel

10 Use the right angle symbol to identify all the right angles on these shapes.

a   b  c

How many right angles does a square have?

11 Add symbols to these shapes in order to identify features such as
sides of equal length, parallel sides and right angles.

a   Rhombus   b   Trapezium   c   Right angle triangle

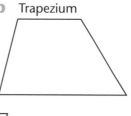

d   Parallelogram   e   Rectangle

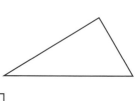

S3.1 Students describe the defining geometric properties of families of 3D shapes, model 3D shapes using nets and other representations, and identify and describe the properties of specific families and subgroups of 2D shapes.

12 Colour each beaker so that they show:

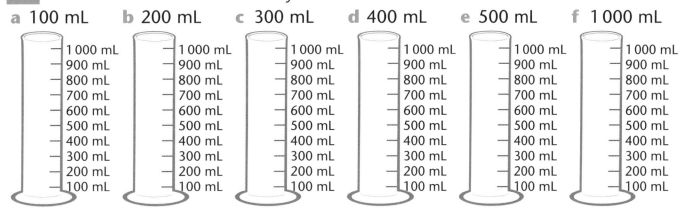

a 100 mL   **b** 200 mL   **c** 300 mL   **d** 400 mL   **e** 500 mL   **f** 1 000 mL

13 Gather these containers, fill them with water, pour the water into beakers and then colour each measuring beaker below to match the quantity of water in each container.

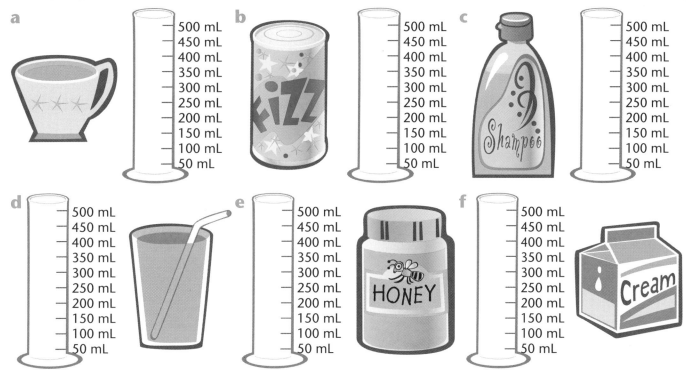

14 Solve these problems.

	Question	Answer
a	What is the difference in capacity between the cream and the shampoo?	mL
b	How many millilitres would two shampoo bottles hold?	mL
c	What is the difference in capacity between the cream and the soft drink?	mL
d	How many millilitres would two cream cartons hold?	mL

M3.1 Students identify and use equivalent forms of standard units when measuring, comparing and ordering, and estimate using a range of personal referents.

63

Thou	Hund	Tens	Ones
	6	12	14
8	~~7~~	~~3~~	~~4~~
− 2	1	8	6
6	5	4	8

8 thousands take away 2 thousands leaves 6 thousands.

6 hundreds take away 1 hundred leaves 5 hundreds.

4 ones take away 6 ones. Trade 1 ten from the tens column to make 14 ones. 14 − 6 = 8

2 tens take away 8 tens. Trade 10 tens from the hundreds column to give 12 tens. 12 tens − 8 tens = 4 tens.

The theatre holds 8 734 seats. How many seats are left if 2 186 tickets have been sold?

1 Complete these subtractions. They do not need regrouping.

a
```
  9 5 6 8
− 4 3 2 5
─────────
```

b
```
  7 8 6 9
− 5 2 3 4
─────────
```

c
```
  8 5 9 6
− 7 1 6 4
─────────
```

d
```
  8 8 9 5
− 5 3 4 2
─────────
```

e
```
  7 9 5 8
− 4 7 4 7
─────────
```

2 Complete the subtractions with regrouping in the ones.

a
```
  7 4 8 5
− 2 4 3 9
─────────
```

b
```
  8 4 7 7
− 3 0 4 9
─────────
```

c
```
  8 4 5 7
− 6 2 3 8
─────────
```

d
```
  7 4 8 5
− 6 3 6 9
─────────
```

e
```
  6 8 4 3
− 2 1 2 8
─────────
```

3 Use regrouping to complete these subtractions.

a
```
  8 3 9 6
− 2 7 4 3
─────────
```

b
```
  7 6 2 9
− 4 1 9 3
─────────
```

c
```
  7 8 5 4
− 3 6 2 8
─────────
```

d
```
  9 5 7 6
− 3 6 2 8
─────────
```

e
```
  5 7 5 4
− 4 9 2 9
─────────
```

I can check my subtractions by backtracking.

```
  7 3 9 2
− 2 7 4 9
─────────
```
⟹
```
    6 13  8 12
  7 ~~6~~ ~~9~~ ~~2~~
− 2 7  4  9
───────────
  4 6  4  3
```
⟹
```
    1       1
  4 6 4 3
+ 2 7 4 9
─────────
  7 3 9 2
```

4 Solve these problems and check your answers using an inverse operation.

a	Sam had saved $9 526 but spent $3 471 on a car. How much money has he left?		c	If 6 289 tickets were sold and 3 092 were sold to children, how many adults bought tickets?	
b	The area of the park is 8 737 square metres but 2 642 square metres has been closed off. How much space is left?		d	The Little Red Car Shop sold 5 215 cars last year but only 3 987 this year. How many more were sold last year?	

N3.2 Students identify and solve addition and subtraction problems involving whole numbers, and decimal fractions in context, selecting from a range of computation methods, strategies and known number facts.   PA 3.1

Counting forwards and backwards

5 Continue the counting patterns.

a	120	130	140		
b	310	300	290		
c	230	270	310		
d	340	290	240		
e	760	660	560		

f	400	600	800		
g	890	790	690		
h	20	520	1020		
i	1350	1450	1550		
j	1260	2260	3260		

6 Follow the counting instructions to complete the patterns.

a Count forwards by 10s

30				

b Count forwards by 100s

200				

c Count forwards by 10s

166				

d Count backwards by 10s

284				

e Count backwards by 100s

986				

f Count forwards by 200s

3 313				

g Count backwards by 50s

5 428				

h Count forwards by 1 000s

1 526				

i Count forwards by 500s

738				

j Count backwards by 100s

9 025				

7 Peta has created some patterns but they are not all correct. Place a tick at the end of each pattern if it is correct. If you think it has an error circle the incorrect number.

a Doubling

3	6	12	24	48	86	172	344

b Backwards by 100

989	789	689	589	489	389	289	189

c Forwards by 20

1770	1790	1810	1820	1840	1860	1880	1900

Captain Cook landed in Australia in 1770
ENDEAVOUR

Organising data by classifying

Taylor's class collected some garbage from around the school. They worked out that the garbage could fit into 5 categories and could be recycled.

8 Sort the garbage into categories by colouring a bin for each piece of garbage. The plastic category has been done for you.

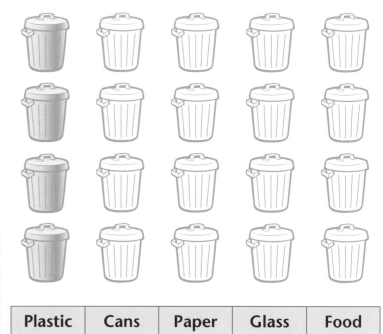

Plastic	Cans	Paper	Glass	Food

9 Group the items into 4 categories. Give each category a name and list all its members in the same box.

red	green	Maths	football	pink	Jupiter
Earth	soccer	Saturn	yellow	Uranus	softball
Mars	Venus	hockey	English	cricket	purple

Name	Colours

Name	

Name	

Name	

10 Compare your categories with other groups in the class and discuss how your categories are similar or different.

CD3.2 Students design and trial a variety of data collection methods and use existing sources of data to investigate their own and others' questions, organise data and create suitable displays identifying and interpreting elements of the displays.

Grams

11 Estimate then measure the mass of each item in grams using a pan balance.

a

Est: _____

Mass: _____

b

Est: _____

Mass: _____

c

Est: _____

Mass: _____

d

Est: _____

Mass: _____

e

Est: _____

Mass: _____

f

Est: _____

Mass: _____

g

Est: _____

Mass: _____

h

Est: _____

Mass: _____

12 What would be the mass of your pencil case with these things in it? You may need a calculator to work it out.

Calculator	
Pencil case	
Scissors	
Pen	
Eraser	
Total	

13 Refer to a set of kitchen scales to work out how many grams each of these measurements represents.

a 1.5 kg = _____ grams   **f** 2.3 kg = _____ grams

b 6.5 kg = _____ grams   **g** 2.7 kg = _____ grams

c 2.5 kg = _____ grams   **h** 3.5 kg = _____ grams

d 3.7 kg = _____ grams   **i** 1.8 kg = _____ grams

e 2.6 kg = _____ grams   **j** 2.1 kg = _____ grams

M3.1 Students identify and use equivalent forms of standard units when measuring, comparing and ordering, and estimate using a range of personal referents.

67

Tenths as decimal fractions

One tenth is equal to 10 hundreths and can be written as 0.1 or 0.10.

1 On the number line below, fill in the empty boxes to show the matching decimal fraction for each common fraction given.

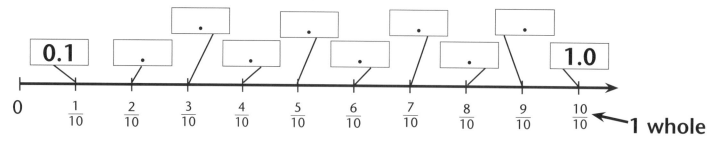

2 Connect each circled decimal to its position on the number line.

a

(0.4)   (0.7)   (1.2)   (1.8)

b

(2.2)   (2.6)   (2.9)   (3.1)   (3.4)   (3.9)

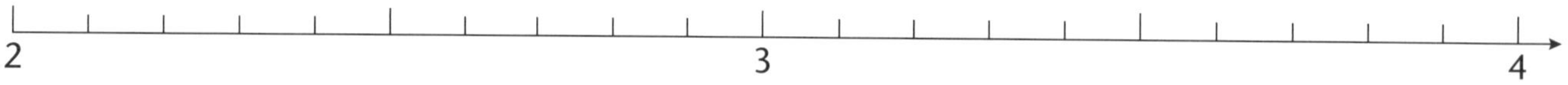

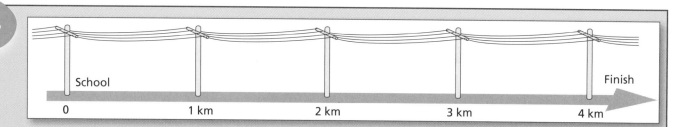

3

Five children are on a walkathon. Put a cross between the electricity poles to estimate where each child is.

a Gavin 0.5 km   **b** Sarah 1.8 km   **c** Bill 2.4 km   **d** Jade 3.9 km

4 Complete these patterns on your calculator:

Rule: 2 + 0.1 = = = =

2	2.1	2.2						

Rule: 5 + 0.5 = = = =

5	5.5	6						

Rule: 10 − 0.2 = = = =

10	9.8	9.6						

N3.1 Students compare, order and represent whole numbers to 9 999 and common and decimal fractions, calculate cash transactions and describe other methods of payment.

PA 3.1

Representing relationships

5 Display the answers to each problem in the tables below and state the rule you used.

a Kris is paid $5 per hour for his job as a paper boy. How much will he earn in 6 hours?

Hours	1	2	3	4	5	6
Pay	5	10	15			

Rule:

b The tap leaks at a rate of 3 litres per hour. How much water has been wasted in 6 hours?

Hours	1	2	3	4	5	6
Litres						

Rule:

c Potatoes are put into 4-kg bags. How many kilograms of potatoes are in 6 bags?

Bags	1	2	3	4	5	6
Kilograms						

Rule:

d The triathlete runs 6 km every hour in her training session. How far would she run in 6 sessions?

Hours	1	2	3	4	5	6
Kilometres						

Rule:

e Tomatoes are delivered in 7-kg boxes. How many kilograms would there be in 6 boxes?

Boxes	1	2	3	4	5	6
Kilograms						

Rule:

f Each sports booklet has 8 pages. How many pages would there be in 6 booklets?

Booklets	1	2	3	4	5	6
Pages						

Rule:

g Samantha is paid $10 per hour and Tom is paid $9 per hour. How much will they each earn if they work 6 hours?

Hours	1	2	3	4	5	6
Tom's pay						
Sam's pay						

Rule:

Three-dimensional objects can be viewed from the top, front, side and back.

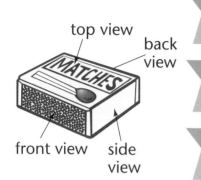

top view

back view

front view   side view

The **top view** only shows what can be seen directly from the top.

The **side view** only shows what can be seen directly from the side.

The **front view** only shows what can be seen directly from the front.

The **back view** is usually the reverse of the front view.

6 Label the views of the objects as top, front, side or back views.

a

_____ view	_____ view	_____ view	_____ view

b

_____ view	_____ view	_____ view	_____ view

c

_____ view	_____ view	_____ view	_____ view

7 Construct the prisms made from blocks then draw the top, front, side and back views of each one.

Front view	Side view	Top view	Back view

a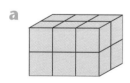

b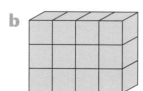

 S3.1 Students describe the defining geometric properties of families of 3D shapes, model 3D shapes using nets and other representations, and identify and describe the properties of specific families and subgroups of 2D shapes.

Minutes to the hour

Minutes past the hour

20 to 2

8 Use the large clock-face to help you write the analogue times.

a
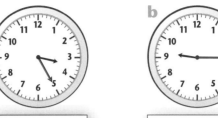

past

b

past

c

to 2

d

e

f

g

I'll meet you at 10 past 4.

h

i

j

k

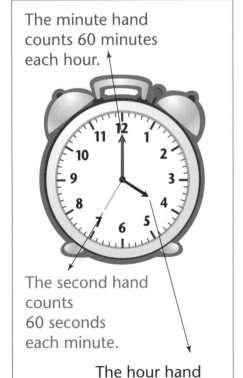

The minute hand counts 60 minutes each hour.

The second hand counts 60 seconds each minute.

The hour hand counts the hours.

9 How many minutes does it take for the minute hand to:

a move from the 3 to the 4? _____

b move from 12 to 6? _____

c make a full revolution of the clock-face? _____

10 How many hours does it take the hour hand to:

a move from one number to the next? _____

b move from the 6 to the 9? _____

c make a full revolution of the clock-face? _____

11 How many minutes does it take the hour hand to move from the 3 to the 4? _____

12 How many seconds does it take the second hand to:

a move from the 12 to the 1? _____

b move from the 12 to the 3? _____

c make a full revolution of the clock-face? _____

M3.2 Students read, record and calculate with 12-hour time, and interpret calendars and simple timetables related to daily activities.

71

4-digit addition

1 Add the 4-digit numbers with trading in the ones.

a	4 2 3 8	b	4 3 2 5	c	7 5 2 6	d	3 7 2 6	e	3 5 0 5
+	2 7 2 4	+	3 5 6 5	+	1 3 5 7	+	5 2 3 8	+	4 3 2 7

2 Add the 4-digit numbers with trading in the ones or tens.

a	7 3 5 7	b	1 0 8 4	c	3 4 8 3	d	6 6 0 7	e	5 8 9 7
+	2 1 2 6	+	3 6 5	+	2 3 4 1	+	4 0 6	+	7 8

3 Add the 4-digit numbers with trading in the ones, tens or hundreds.

a	3 5 7 4	b	6 8 3 6	c	3 5 7 9	d	2 6 7 4	e	3 5 8 6
+	2 3 8 0	+	1 7 0 6	+	3 5 5 0	+	4 6 3 8	+	1 5 5 7

4 How much did each person spend? (You may need a calculator.)

Entrée		*Main*		*Dessert*	
Prawns	$6.50	Fish	$8.50	Apple pie	$2.50
Oysters	$7.50	Fillet steak	$12.00	Ice cream	$1.50
Soup	$5.50	Chicken	$9.50	Strawberries	$4.50
Mushrooms	$6.00	Lobster	$21.00	Pavlova	$3.50
		Steak diane	$12.50		
		Chicken nuggets	$6.50		
		Hamburger	$7.00		

a Mr Barton had an entrée of prawns, a main meal of
fish and an apple pie for dessert. $ _____

b Mrs Hill had oysters, steak diane and pavlova. $ _____

5 Use strategies such as 'guess and check' to make up a meal consisting of an
entrée, main and dessert that costs
between $23 and $26.

Food items	Cost
Total cost	

N3.2 Students identify and solve addition and subtraction problems involving whole numbers, and decimal
fractions in context, selecting from a range of computation methods, strategies and known number facts.   **PA3.2**

6 Fibonacci numbers

Leonardo Fibonacci of Pisa in Italy discovered this number pattern.

a Look at the pattern and write the next 7 Fibonacci numbers.

1, 1, 2, 3, 5, 8, 13, ___, ___, ___, ___, ___, ___, ___,

Leaning Tower of Pisa

b Explain how the rule above works.

_____

_____

7 Study the timetable to see if there are any patterns.

Station	Train					
	1	2	3	4	5	6
	a.m.	a.m.	a.m.	a.m.	a.m.	a.m.
Abbottville	9:05	9:25	9:45	10:05	10:25	10:45
Bentley	9:15	9:35	9:55	10:15	10:35	10:55
Cartertown	9:25	9:45	10:05	10:25	10:45	11:05
Dunton	9:35	9:55	10:15	10:35	10:55	11:15
Fairview	9:45	10:05	10:25	10:45	11:05	11:25

a What pattern could you see as you looked across the timetable from Train 1 through to Train 6?

_____

b What pattern could you see as you looked down the timetable, that is, as the train travelled from Abbottville to Fairview?

_____

c Describe any other pattern you can see.

_____

8 Follow the rules to complete the patterns.

Add 70

a
75	85	95	105	115	125
145	155				

Subtract 30

c
780	680	580	480	380	280
750					

Add 10

b
1.1	1.2	1.3	1.4	1.5	1.6
11.1	11.2				

Double

d
2	20	200	2 000	20 000	200 000

9 Flip, slide or turn the following shapes.
Draw your results.

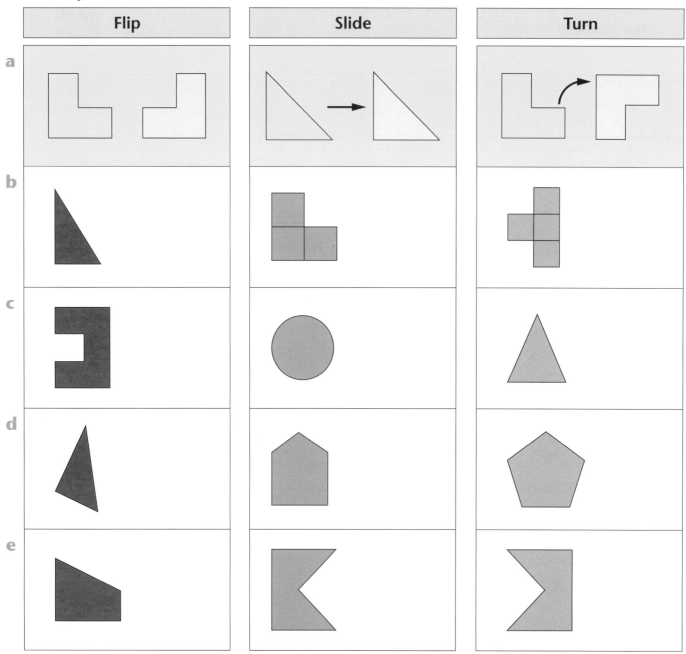

	Flip	Slide	Turn
a			
b			
c			
d			
e			

10 Help Wally move the box along the path by continuing his pattern of moves.

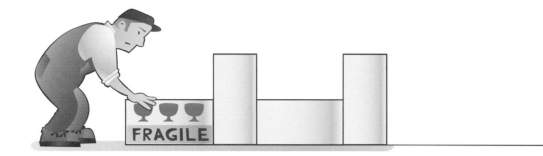

FRAGILE

S3.1 Students describe the defining geometric properties of families of 3D shapes, model 3D shapes using nets and other representations, and identify and describe the properties of specific families and subgroups of 2D shapes.

11 Which colour is most likely and least likely to occur?

a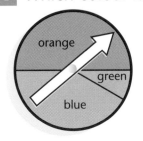

b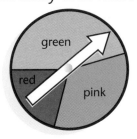

c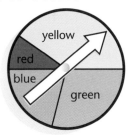

d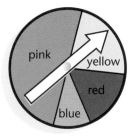

Most likely

Most likely

Most likely

Most likely

_____

_____

_____

_____

Least likely

Least likely

Least likely

Least likely

_____

_____

_____

_____

e Design a spinner where pink is most likely and green is least likely to occur.

f Design a spinner where green is the most likely and pink and blue have the same chance of occurring.

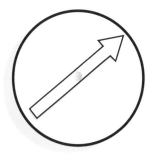

12 Answer True or False to describe these events.

Event	T or F
a It's more likely that Mum will get home before I do today.	
b It's more likely that I'll bring lunch from home rather than buy it from the canteen.	
c It's more likely that 2 people from our class will receive awards at the assembly than 10 people.	
d It's less likely that we'll get homework on Friday than on Monday.	
e It's less likely that our teacher will walk to school than drive.	
f If my name was put in a hat with ten other names it is equally likely that my name will be selected.	
g It is equally likely that I will throw a six to start a game as any other number.	

13 Name an event that has two possible results, each with an equal chance of happening.

CD3.1 Students identify all possible outcomes of familiar situations or actions and, for these sample spaces, order the likelihood of occurrence of the identified outcomes using experimental data.

75

Diagnostic review 2

PART 1

Complete the additions and subtractions.

a
```
  3 5 7 4
+ 2 3 1 7
---------
```

b
```
  7 5 3 8
+ 1 3 9 6
---------
```

c
```
  8 3 7 2
- 2 4 3 6
---------
```

d
```
  6 3 5 4
- 4 2 7 6
---------
```

Estimate an answer to the following by rounding off to the nearest 1000.

e	4 087 + 4 968	
f	2 107 + 3 945	
g	1 201 + 1 299	

PART 2   N3.3

Use strategies such as doubling to solve these.

a 7 × 4 = ☐   b 14 × 4 = ☐

c 40 × 4 = ☐   d 18 × 4 = ☐

e 12 × 6 = ☐   f 9 × 6 = ☐

Use any strategy you wish to solve these.

g 16 × 5 = ☐   h 14 × 5 = ☐

i 18 × 5 = ☐   j 9 × 5 = ☐

Use the 'double, double and double again' strategy to solve these.

k 9 × 8 = ☐   l 16 × 8 = ☐

m 12 × 8 = ☐   n 20 × 8 = ☐

List the next 6 multiples of the given numbers.

o	8						
p	4						
q	9						

PART 3   N3.3

Complete these division facts.

a 18 ÷ 6 = ☐   b 24 ÷ 6 = ☐

c 40 ÷ 8 = ☐   d 35 ÷ 7 = ☐

e 30 ÷ 6 = ☐   f 48 ÷ 6 = ☐

PART 4   N3.1

Continue the pattern of mixed numbers.

a $0, \frac{1}{4}, \frac{2}{4}, \frac{3}{4}, 1, 1\frac{1}{4}$ ___ ___ ___

b $\frac{3}{5}, \frac{4}{5}, 1, 1\frac{1}{5}$ ___ ___ ___ ___

c $2\frac{1}{4}, 2\frac{2}{4}, 2\frac{3}{4}, 3$ ___ ___ ___

PART 5   N3.1

Shade each shape to represent the decimal below it.

a [ ] 0.8   b [ ] 0.2

c Order the decimals from smallest to largest: 0.6 0.4 0.8 0.9

_____

PART 6   PA3.1

Complete the tables to solve the problems.

a William scored 5 tries in the rugby game. What was the total number of points he scored?

Tries	1	2	3	4	5
Points	5	10			

b Nathan earns $12 per hour. How much will he earn for 5 hours work?

Hours	1	2	3	4	5
Dollars	$12				

Diagnostic review 2

Part 7
S3.1

Draw the 3D objects. Include edges you can't see by dotting the lines.

a   b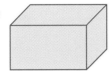

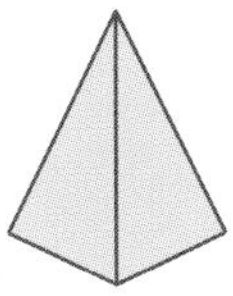

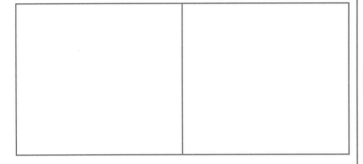

Part 8
S3.1

Use arrows to identify all parallel lines and identify all right angles and sides of equal length using symbols.

Part 9
S3.1

Use the grids to show the top, front, side and back views of the shape.

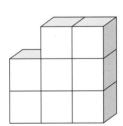

Top		Front	

Side		Back	

Part 10
M3.1

Colour the beaker to show the volume of the containers.

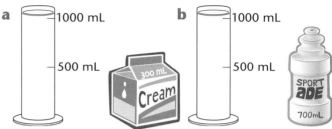

a 1000 mL / 500 mL   b 1000 mL / 500 mL

Part 11
M3.1

Name one item that matches each category.

About 100g	About 200g	About 500g

Part 12
M3.1

Work out the area of this shape in square centimetres.

[ ] square centimetres

Part 13
M3.1

Draw lines to connect each item with its mass.

200 g	500 g	1 kg

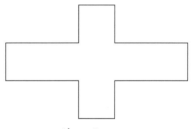

Part 14
CD3.1

a Which colour is more likely to be spun? _____

b Which colour is least likely to be spun? _____

c Does yellow have a greater chance of being spun than orange? _____

77

Inverse operations

1 Sarah answered 4 questions in a test. Test her accuracy by either using the inverse operation or repeating the same operation in the space below her work.

a

Thou	Hund	Tens	Ones
3	4	5	6
+ 2	3	3	6
5	7	9	2

Thou	Hund	Tens	Ones

_____

_____

b

Thou	Hund	Tens	Ones
6	7	3	2
+ 2	5	1	6
8	2	4	8

Thou	Hund	Tens	Ones

_____

_____

c

Thou	Hund	Tens	Ones
4	2	7	4
− 2	6	6	5
1	6	1	9

Thou	Hund	Tens	Ones

_____

_____

d

Thou	Hund	Tens	Ones
8	6	0	0
− 2	4	3	4
6	1	6	6

Thou	Hund	Tens	Ones

_____

_____

$$\begin{array}{r} 5\,7\,9\,2 \\ -\,2\,3\,3\,6 \\ \hline 3\,4\,5\,6 \end{array}$$

2 Explain what Sarah did wrong in question **c** above.

$$\begin{array}{r} 4\ \ 2\ \ 7\ \ 4 \\ -\ 2\ \ 6\ \ 6\ \ 5 \\ \hline \end{array}$$

_____

_____

Always check your work.

3 Solve these problems and check your answers using another method.

a	Gina saved $4 278 in May and $2 429 in June. What was the total amount she saved over the two months?		
b	The container originally held 5 500 mL. How much is left if Jessica used 1 255 mL to fertilise her plants?		
c	Out of the 9 520 tickets that went on sale today there are only 1 256 tickets left. How many have been sold?		
d	If Kim has $4 578 but still needs another $2 349 to buy the car she wants, how much is the car?		
e	The total mass of the container was 8 844 kg, what would its mass be if the model train with a mass of 1 560 kg was taken out?		

N3.2 Students identify and solve addition and subtraction problems involving whole numbers, and decimal fractions in context, selecting from a range of computation methods, strategies and known number facts.

PA 3.1

4 Reduce each group so that it is $\frac{1}{2}$ its original size. Circle those that will remain.

a

b

c

$10 \times \frac{1}{2} = \boxed{}$

$18 \times \frac{1}{2} = \boxed{}$

$12 \times \frac{1}{2} = \boxed{}$

5 Draw extra items to make each group $1\frac{1}{2}$ times its original size.

a

b

$4 \times 1\frac{1}{2} = \boxed{}$

$6 \times 1\frac{1}{2} = \boxed{}$

$8 \times 1\frac{1}{2} = \boxed{}$

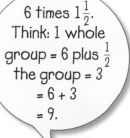

6 times $1\frac{1}{2}$.
Think: 1 whole group = 6 plus $\frac{1}{2}$ the group = 3
= 6 + 3
= 9.

c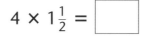

6 Kelly has a herd of 24 cattle.

a How many cows are in Sam's herd if it is $\frac{1}{2}$ the size of Kelly's?

_____

b How many cows are in Con's herd if it is $1\frac{1}{2}$ times the size of Kelly's?

_____

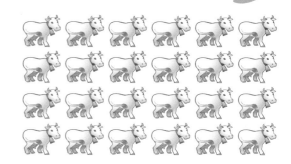

7 Complete these patterns.

Rule: halve

a

	120	110	100	90	80	70	60	50
	60	55					30	

Rule: $1\frac{1}{2}$

b

	10	12	14	16	18	20	22	24
	15			24				36

8 How many:

a minutes in $1\frac{1}{2}$ hours?

_____

b grams in $1\frac{1}{2}$ kg?

_____

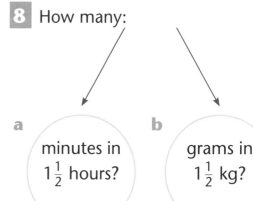

N3.3 Students identify and solve multiplication and division problems involving whole numbers, and decimal fractions in context, selecting from a range of computation methods, strategies and known number facts.

79

9 Use the terms **flip**, **slide** and **turn** to describe the movement of the triangle.

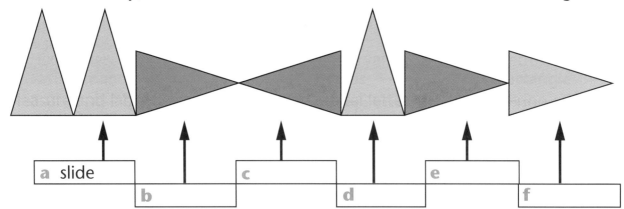

a slide

b

c

d

e

f

10 Follow the instructions to make a pattern.

a Flip the shape repeatedly.

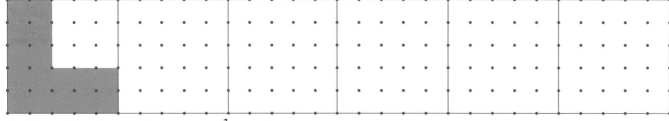

b Turn 90° in each square ($\frac{1}{4}$ turn).

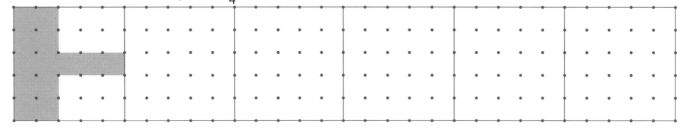

11 Make a pattern of your own.

12 Follow the instructions to make a pattern.

flip   turn 90°   flip   turn 90°   flip

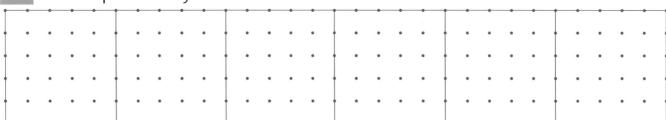

S3.1 Students describe the defining geometric properties of families of 3D shapes, model 3D shapes using nets and other representations, and identify and describe the properties of specific families and subgroups of 2D shapes.

The square metre

A square centimetre is too small to measure large areas. Larger areas need to be measured in **square metres**.

1 m

1 m

1 square metre

13 Make three areas which are all equal to one square metre.

- The first one will be a square with all sides 100 cm long.
- The second one will be a rectangle 200 cm long and 50 cm wide.
- The third one will be a rectangle 400 cm long and 25 cm wide.

100 cm

100 cm

200 cm

50 cm

400 cm

25 cm

14 Explain why all these shapes have an area of one square metre.

_____

_____

15 Use your newspaper square metres to identify areas in the playground that are less than 1 square metre, about 1 square metre and larger than 1 square metre.

Less than 1 square metre	About 1 square metre	Larger than 1 square metre

16 Estimate and then measure these areas using the square metres you and your classmates have made.

	Estimate	Square metres
Chalkboard		
Classroom floor		
Library floor		
Classroom window		
Noticeboard		

M3.1 Students identify and use equivalent forms of standard units when measuring, comparing and ordering, and estimate using a range of personal referents.

81

Division with remainders

Question: What would happen if 30 was divided by 4?
Answer: There would be 7 groups of 4 and a **remainder** of 2, because
7 x 4 = 28 and 2 more makes 30.

1 Use multiplication and division facts to solve the number sentences.

a 13 ÷ 4 = ☐ remainder ☐   f 21 ÷ 4 = ☐ remainder ☐

b 10 ÷ 3 = ☐ remainder ☐   g 22 ÷ 5 = ☐ remainder ☐

c 13 ÷ 3 = ☐ remainder ☐   h 27 ÷ 5 = ☐ remainder ☐

d 12 ÷ 5 = ☐ remainder ☐   i 26 ÷ 4 = ☐ remainder ☐

e 16 ÷ 5 = ☐ remainder ☐   j 16 ÷ 3 = ☐ remainder ☐

2 Solve these 'division with remainders' problems. Firstly make the shares equal and
then explain what you would do with the remainders. The first one is done for you.

	Problem	Division fact with remainder	Solution
a	Farmer Ted has 28 sheep which are to be put in 3 paddocks. How many in each?	9 × 3 = 27 and 1 more makes 28.	Place the remaining sheep in a paddock to make 1 group of 10 and 2 groups of 9.
b	44 marbles were shared among 6 children. How many did each child receive?		
c	48 dice were shared among 5 groups in the classroom. How many did each group receive?		
d	23 football cards were shared among 4 girls. How many did each girl receive?		
e	45 cows were grouped in 8s for sale. How many groups were there?		
f	50 flowers were planted in 6 gardens. How many flowers in each garden?		

3 Explain the strategy you would use to solve 32 ÷ 5 = ☐

_____

_____

_____

N3.3 Students identify and solve multiplication and division problems involving whole numbers, and decimal fractions in context, selecting from a range of computation methods, strategies and known number facts.

Decimal fractions

Decimal notation

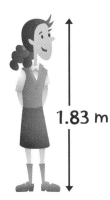

1.83 m

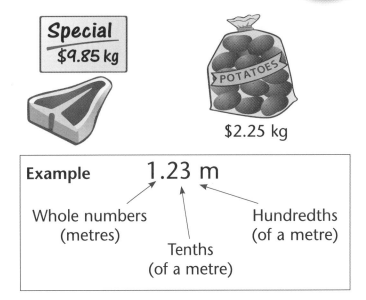

Special
$9.85 kg

POTATOES

$2.25 kg

Example	1.23 m

Whole numbers (metres)

Tenths (of a metre)

Hundredths (of a metre)

Each of these measurements has a **decimal point**. The decimal point separates the fractional part of a number from the whole.

---

4 Measure the heights of 9 volunteers in your class, carefully completing the table as you go. One example has been done for you.

	Name	Centimetres	Metres and centimetres	Decimal
a	Harriet	183 cm	1 m and 83 cm	1.83 m
b				
c				
d				
e				
f				
g				
h				
i				
j				

5 Six children from Ms Gem's class entered the high jump and recorded their scores.

Maria	1.12 m	Moana	0.69 m	Solomon	0.93 m
Trent	0.72 m	Nathan	1.04 m	James	0.89 m

Order the jumps from smallest to largest. (Measure them if you need to.) _____

_____

6 What fruit is the most popular in your class? Predict answers to the following questions about the most popular fruits in your class.

An apple a day!
The Queensland Government is urging school canteens to sell more nutritious foods. Fruit is …

	Prediction	Survey result
The most popular fruit will be		
The least liked fruit will be		

7 Conduct a survey to find out which are the 5 most popular fruits, then tally and graph your results in the space below.

___apples___ _____ _____ _____ _____

Most popular fruits, Class _____

Fruits		
apples		

1 2 3 4 5 6 7 8 9 10 11 12 13 14 15
Number of children

8 In groups, repeat your survey using a different class or group from within your school.

9 Using all the data you have collected, suggest the types of fruit that Queensland canteens should sell. _____

CD3.2 Students design and trial a variety of data collection methods and use existing sources of data to investigate their own and others' questions, organise data and create suitable displays identifying and interpreting elements of the displays.

Metres and centimetres

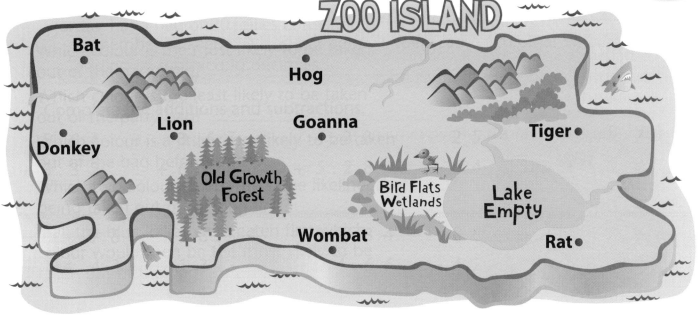

ZOO ISLAND

10 Landell has drawn a plan of Zoo Island. Measure the distances to the nearest centimetre on her map in order to answer these questions.

a How far is it from Bat to Hog? _____

b How far is it from Donkey to Lion? _____

c How far is it from Bat to Wombat in a straight line? _____

d How far is it from Rat to Tiger? _____

e How far is it from Lion to Tiger? _____

f How wide is Lake Empty? _____

11 Draw lines to connect Lion, Goanna and Wombat. Colour this area.

a What shape is this area? _____

b What is its perimeter? _____

12 Draw lines to connect Wombat, Rat, Tiger and Goanna. Colour this area.

a What shape is this area? _____

b What is its perimeter? _____

13 Add these towns to the map.

a Monkey: 4 cm west of Wombat

b Duck: 6 cm east of Hog

14 Below are the 5 best javelin throws recorded at the athletics carnival. Convert them from decimal fractions to centimetres and place the jumps in order from 1st to 5th.

	Name	Metres	Centimetres	Place
a	Serena	15.42		
b	James	12.38		
c	Ronald	16.27		
d	Kirsten	19.95		1st
e	Ellen	18.08		

M3.1 Students identify and use equivalent forms of standard units when measuring, comparing and ordering, and estimate using a range of personal referents.

85

Division strategies

Jimmy wants to share 22 marbles into 5 bags. He knows that it will not work out equally and that there will be a **remainder**.

4 r 2
5⟌22

22 shared among 5 bags means 4 in each bag and 2 left over.

1 Solve these divisions that have remainders.

a 3⟌1 3	**e** 3⟌1 9	**i** 7⟌2 2	**m** 8⟌2 5	Use multiplication facts to help you.
b 4⟌1 7	**f** 4⟌2 2	**j** 6⟌1 9	**n** 9⟌3 0	
c 5⟌2 1	**g** 5⟌3 6	**k** 7⟌3 0	**o** 8⟌3 5	
d 6⟌2 5	**h** 6⟌3 1	**l** 8⟌3 3	**p** 7⟌5 0	

2 Write down the division fact that helps you solve the questions below and briefly explain what you would do with the remainder.

Example: 25 children are to be placed into 4 teams. The division fact I know is $24 \div 4 = 6$. My solution would be 3 teams of 6 and one team of 7.

	Question	Division fact	My solution
a	29 people into 3 teams		
b	19 cards into 4 groups		
c	42 girls into 8 teams		
d	38 beads on 5 bangles		
e	31 people into 6 buses		
f	46 people into 9 groups		
g	64 cows into 8 paddocks		

3 Solve the problems.

a	Jim had 45 football cards that he shared equally among 5 children. How many did each child receive?		**c**	Sally bought 6 CDs at the music store. What was the cost of each CD if the total cost for all of them was $54 and each CD was the same price?	
b	Mr Malouf put 48 children into 8 equal teams. How many were in each team?		**d**	Jack had 48 marbles that he shared between himself and 3 other children. How many marbles did each child receive?	

N3.3 Students identify and solve multiplication and division problems involving whole numbers, and decimal fractions in context, selecting from a range of computation methods, strategies and known number facts.

4 Study the grids then complete the table.

Concrete	$\frac{\square}{10}$	$\frac{\square}{100}$	Dec.
a	$\frac{3}{10}$	$\frac{30}{100}$	0.30
b			
c		$\frac{60}{100}$	
d			0.80

Concrete	$\frac{\square}{10}$	$\frac{\square}{100}$	Dec.
e		$\frac{40}{100}$	
f			0.70
g	$\frac{5}{10}$		
h			0.9

5 Use the greater than >, less than <, or equal sign = to make these number sentences true.

a $\frac{4}{10}$ ☐ $\frac{6}{10}$

b $\frac{70}{100}$ ☐ $\frac{30}{100}$

c $\frac{50}{100}$ ☐ $\frac{40}{100}$

d $\frac{8}{10}$ ☐ $\frac{7}{10}$

e $\frac{8}{10}$ ☐ $\frac{60}{100}$

f $\frac{3}{10}$ ☐ $\frac{30}{100}$

g $\frac{50}{100}$ ☐ $\frac{4}{10}$

h $\frac{4}{10}$ ☐ $\frac{40}{100}$

i $\frac{7}{10}$ ☐ $\frac{50}{100}$

j $\frac{80}{100}$ ☐ $\frac{90}{100}$

k $\frac{4}{10}$ ☐ 0.60

l $\frac{9}{10}$ ☐ $\frac{90}{100}$

m 0.30 ☐ $\frac{4}{10}$

n 0.60 ☐ $\frac{40}{100}$

o $\frac{50}{100}$ ☐ 0.60

$\frac{7}{10} > \frac{2}{10}$ $\frac{2}{10} > \frac{1}{10}$

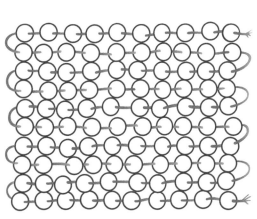

6 Colour the beads to show how Kara coloured them.

a She coloured 1 tenth of them red.

b She coloured 3 tenths of them yellow.

c She coloured 20 hundredths green.

d She coloured 0.30 of them blue.

e She coloured 7 hundredths orange.

f She left 3 hundredths uncoloured.

N3.1 Students compare, order and represent whole numbers to 9 999 and common and decimal fractions, calculate cash transactions and describe other methods of payment.

87

A **triangle** is a 3-sided shape with 3 angles. There are 3 main types of triangles: **equilateral**, **isosceles** and **scalene**. Symbols such as strokes are used to show sides of equal length, and angles that are the same size can be identified using arcs.

7 Study the 3 types of triangles pictured, then answer the questions.

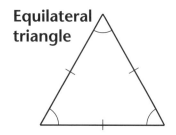
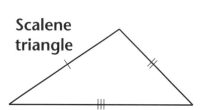
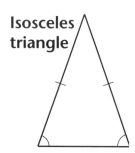

a Which triangle has all sides of equal length? _____

b Which triangle has two sides of equal length? _____

c Which triangle has all angles the same size? _____

d Which triangle has two angles the same size? _____

e Which triangle has no sides the same length? _____

f Which triangle has no angles the same size? _____

A right angle triangle is a triangle in which one angle is a right angle.

8 Colour the right angle triangles.

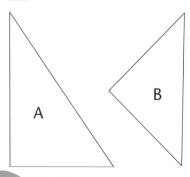
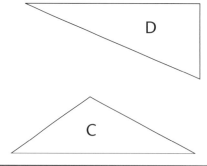
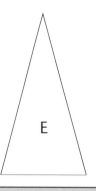

Did you find 3 right-angled triangles?

9 Use triangles to create a tessellating pattern.

S3.1 Students describe the defining geometric properties of families of 3D shapes, model 3D shapes using nets and other representations, and identify and describe the properties of specific families and subgroups of 2D shapes.

a.m. and p.m. time

a.m. is used to describe all time between midnight and noon.
p.m. is used to describe all time between noon and midnight.

10 Write the digital times in words and explain what the time means. The first one is done for you.

	Digital time	Words	This means that it is:
a	2:45 a.m.	two forty-five a.m.	45 minutes past 2 in the morning.
b	9:15 p.m.		
c	6:29 p.m.		
d	8:48 a.m.		
e	9:21 p.m.		
f	11:35 a.m.		

11 Calculate how long it is until the next hour.

a If the time is ⬚ 3:48 a.m. how many minutes is it to 4 a.m.? _____ minutes

b If the time is ⬚ 9:35 p.m. how many minutes is it to 10 p.m.? _____ minutes

c If the time is ⬚ 6:25 a.m. how many minutes is it to 7 a.m.? _____ minutes

d If the time is ⬚ 10:05 a.m. how many minutes is it to 11 a.m.? _____ minutes

e If the time is ⬚ 4:29 p.m. how many minutes is it to 5 p.m.? _____ minutes

12 Draw the times on the clock-faces then write them in the digital form below.

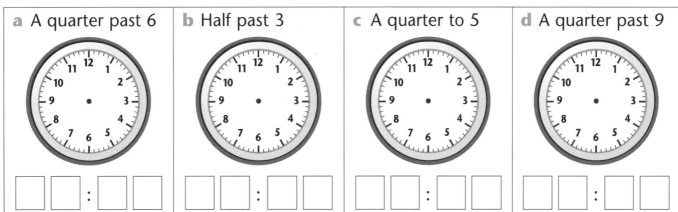

a A quarter past 6   b Half past 3   c A quarter to 5   d A quarter past 9

M3.2 Students read, record and calculate with 12-hour time, and
interpret calendars and simple timetables related to daily activities.

89

Division strategies

Mr Brown had 65 stamps to share between his 5 children. This is what he did.

65 shared among 5.	Share out the tens with each person getting 1 ten.	Trade the 1 ten left for 10 ones. Now share the 15 ones among 5.
5)65	1 5)6⁵5	1 3 5)6⁵5

1 Solve the divisions.

a   2)3 2

b   3)4 2

c   2)5 2

d   4)5 6

e   3)4 5

f   4)5 2

g   5)7 5

h   3)7 5

i   4)6 4

j   6)7 2

k   4)6 8

l   5)8 0

m   8)9 6

n   7)9 1

o   6)8 4

2 Work backwards to check these solutions. Tick the ones you think are correct. The first one has been done for you.

	Problem	Solution	Check
a	19 ÷ 5 =	3 remainder 4	19 − 4 = 15 and 3 × 5 = 15
b	24 ÷ 5 =	4 remainder 4	
c	27 ÷ 4 =	6 remainder 2	
d	19 ÷ 4 =	4 remainder 3	
e	26 ÷ 8 =	3 remainder 2	
f	48 ÷ 9 =	5 remainder 3	
g	95 ÷ 10 =	9 remainder 5	

Take the remainder away. Use a multiplication fact to check the division.

3 Solve these problems.

a | Mr Wilde wants to share 37 books among 5 students. How many will each student receive?

b | Mrs Kelly baked 43 cakes that she shared between 5 children. How many did each child receive?

c | The baker made 24 buns that he wants to put into 4 bags. How many buns will he put in each bag?

d | The zoo has 79 birds that need to be shared between 4 cages. How many birds will there be in each cage?

N3.3 Students identify and solve multiplication and division problems involving whole numbers, and decimal fractions in context, selecting from a range of computation methods, strategies and known number facts.   **PA3.1**

Decimal place value

Decimal notation uses a point to separate the whole numbers from the tenths and hundredths.
Zero can be used as a place holder for hundredths that are less than 10. **E.g. 7 hundredths = 0.07**

| 3 | . | 7 | 4 |

Whole numbers   Tenths   Hundredths

4 If the MAB hundreds block represents one whole, use decimal notation to describe the shaded part of these shapes. The first one is done for you.

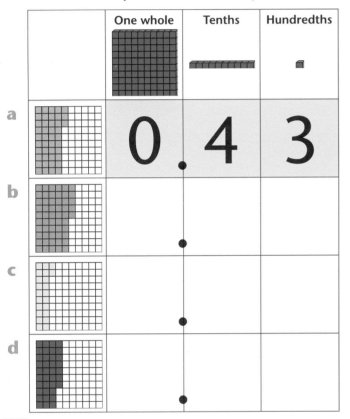

5 Order the decimals from smallest to largest.

a	0.3 m	0.7 m	0.6 m	0.4 m
b	0.6 kg	0.3 kg	0.9 kg	1.0 kg
c	0.63 m	0.27 m	0.94 m	0.16 m
d	$4.32	$4.36	$4.37	$4.39
e	$3.67	$3.36	$3.26	$3.74
f	2.36 m	2.14 m	1.32 m	3.26 m
g	$25.34	$24.93	$42.61	$61.82

Is 0.9 greater than 1?

? ? ?

6 Add any numbers you wish to this group which is arranged smallest to largest.

11.52 m, 11.59 m, ☐ , 13.06 m, ☐ , 15.89 m, ☐ , ☐ , 19.98 m

N3.1 Students compare, order and represent whole numbers to 9 999 and common and decimal fractions, calculate cash transactions and describe other methods of payment.

91

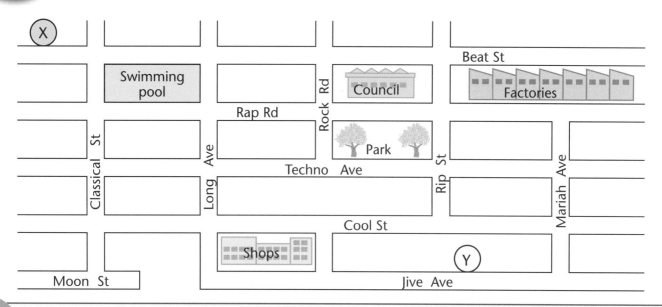

7 Identifying paths

a Draw 3 different paths using 3 different coloured pencils to show how to get from X to Y.

b Which one of the paths was the shortest? _____

c Which one was the longest? _____

d Write a set of instructions that describes one of your paths.

_____

_____

_____

Trent drew a bird's eye view of his lounge room.

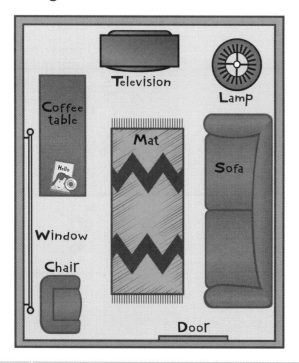

8 Imagine you are looking down on your lounge room from a bird's eye view and draw a plan view of it. You only need to draw the tops of objects.

S3.2 Students interpret and create maps and plans using a range of conventions, describe locations and give directions using major compass points, angles and grids.

Measuring volume

To work out the volume of the shape, calculate how many cubes are in the bottom layer of each model and then multiply that number by the number of layers in each model.

The bottom layer is made up of 4 cubes.
There are 3 layers.
The volume is equal to 3 lots of 4.
The volume = 12 cubes.

9 Build these models using construction cubes. Record the number of cubes in the bottom layer and use this information to calculate the shape's volume.

a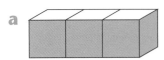

b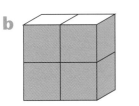

c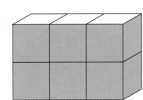

Model	Cubes per layer	Number of layers	Volume
A			
B			
C			
D			
E			
F			

There are 3 cubes in each layer. There are 2 layers. That's 6 cubes.

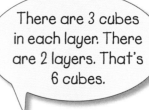

d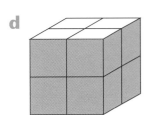

e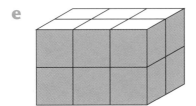

f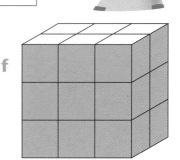

10 Jack has filled a stapler box and a sultana box with cubes in order to measure their volume. Record the number of cubes used in the box below each model.

a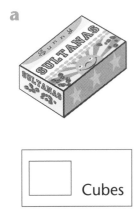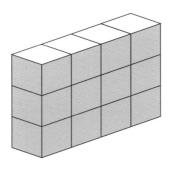

☐ Cubes

b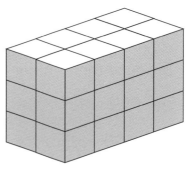

☐ Cubes

M3.1 Students identify and use equivalent forms of standard units when measuring, comparing and ordering, and estimate using a range of personal referents.

93

Factors

Factors are whole numbers that can be multiplied with another number to make a new number. For example: the factors of 16 are 1, 2, 4, 8 and 16.   (2 × 8 = 16   4 × 4 = 16   16 × 1 = 16)

1 Answer true or false.

a 3 is a factor of 6 _____

b 7 is a factor of 15 _____

c 5 is a factor of 20 _____

d 4 is a factor of 13 _____

e 10 is a factor of 50 _____

f 6 is a factor of 18 _____

2 Use division to find the missing factor.

a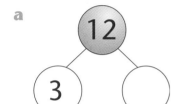
(12) — 3, ○

b (18) — 2, ○

c (24) — 4, ○

d (36) — 9, ○

Multiplication can be made easier by **factorising**. To factorise a number you simply write it as a product of its factors.
E.g. 16 can be written as 4 × 4, 2 × 8 or 16 × 1.

3 Supply another factor for the larger number then complete the multiplications.

a 16 × 5   becomes   8 × 2 × 5 = 80

b 12 × 5   becomes   6 × ☐ × 5 = ☐

c 14 × 5   becomes   7 × ☐ × 5 = ☐

d 18 × 5   becomes   9 × ☐ × 5 = ☐

All multiples of 10 always have 2, 5 and 10 as one of their factors.

☐ × ☐

4 Work backwards to find 3 numbers that multiply together to produce the number in the box.

a   60 = ☐ × ☐ × ☐ =

b   100 = ☐ × ☐ × ☐ =

c   140 = ☐ × ☐ × ☐ =

PA3.1 Students create and continue number patterns, identify, describe and represent relationships between two quantities and use backtracking to reverse any one of the four operations.

N3.3

Rounding decimals and money

When we **round off** decimals to the nearest whole number, the decimals that are .5 or above are rounded up and the decimals below .5 are rounded down. For example, 3.64 is rounded to 4.

5 What whole number would the balls roll to?

a 0.75 _____ b 1.23 _____ c 1.68 _____ d 2.21 _____ e 2.55 _____

6 Round these amounts to the nearest dollar.

a $1.88 = _____   e $1.95 = _____   i $2.99 = _____

b $3.19= _____   f $2.14 = _____   j $7.96 = _____

c $2.94 = _____   g $8.76 = _____   k $3.11 = _____

d $3.82 = _____   h $4.98 = _____   l $2.01 = _____

$7.96 is about $8.

7 Use your rounding strategies to find the approximate totals.

a $3.95 + $2.95 + $4.90 ≈ $ _____   d $6.60 + $5.95 + $4.40 ≈ $ _____

b $8.10 + $6.80 + $4.95 ≈ $ _____   e $9.25 + $9.25 + $9.50 ≈ $ _____

c $7.85 + $2.20 + $2.20 ≈ $ _____   f $8.75 + $5.95 + $4.30 ≈ $ _____

8 Calculate the approximate change each person would receive if:

a Jennifer had $20 but spent $14.95 _____   d Amity spent $49.60 of $100 _____

b Michael had $40 but spent $27.99 _____   e Dale spent $95.99 of $500 _____

c Jessica had $50 but spent $34.75 _____   f Caleb spent $14.99 of $200 _____

9

 Coffee $2.60   Milk $1.30   Jam $3.40   Honey $2.70

Mrs Ramsden is in a huge hurry to get to the staffroom. On her way she has to buy coffee, milk, jam and honey. She only has $10 in her purse.

a Quickly estimate if $10 is enough to buy what she wants. _____

b Explain your strategy. _____

_____

10 **a** Draw new shapes double the size of the ones below.

b Use symbols to identify sides of equal length, angles of equal size and parallel sides.

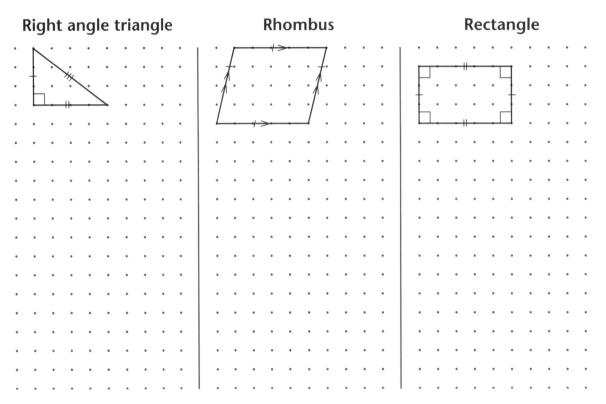

Right angle triangle   Rhombus   Rectangle

11 Compare the properties of your enlarged shapes to the smaller ones.

a Is your triangle a right angle triangle? _____

b Do the other angles on your triangle look to be the same size? _____

c Are all the sides on the small rhombus the same length? _____

d Are all the sides on the large rhombus the same length? _____

e Are the opposite sides on the small rhombus parallel to each other? _____

f Are the opposite sides on the large rhombus parallel to each other? _____

g Are the opposite sides on the small rectangle parallel to each other? _____

h Are the opposite sides on the large rectangle parallel to each other? _____

i What type of angles do rectangles have? _____

12 Draw a trapezium half the size of this one.

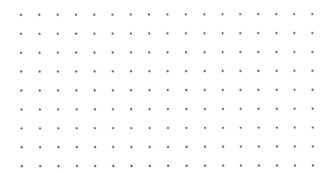

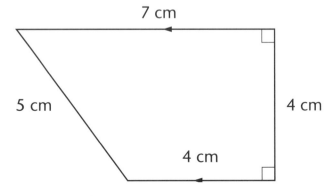

7 cm

5 cm

4 cm

4 cm

S3.1 Students describe the defining geometric properties of families of 3D shapes, model 3D shapes using nets and other representations, and identify and describe the properties of specific families and subgroups of 2D shapes.

Kilograms and grams

13 Read the scales to record the masses in kilograms.

a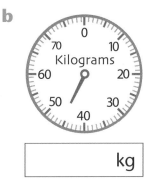

kg

b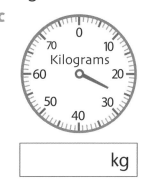

kg

c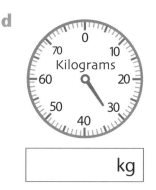

kg

d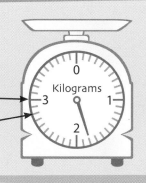

kg

The masses of smaller objects are often measured in **kilograms** and **grams**. Kitchen scales are often used for these smaller masses.
The large numbers on the scales are the whole kilograms.
The smaller markings are steps of 100 grams.

14 Record the mass shown on each set of scales in kilograms and grams and as kilograms using decimal fractions. The first one has been started for you.

a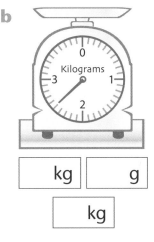

| kg | g |

2.9 kg

b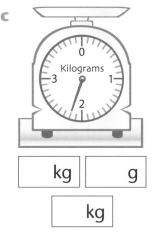

| kg | g |

kg

c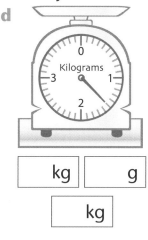

| kg | g |

kg

d

| kg | g |

kg

15 Draw a needle on the scales to show the masses.

a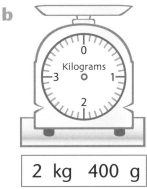

1.3 kg

b

2 kg 400 g

c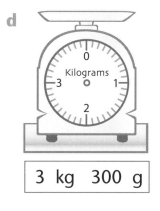

2.9 kg

d

3 kg 300 g

M3.1 Students identify and use equivalent forms of standard units when measuring, comparing and ordering, and estimate using a range of personal referents.

97

Adding and subtracting 4-digit numbers

- Subtract the ones.
  To take 6 from 0 trade
  a ten for 10 ones.
  10 take away 6.
- Subtract the tens.
  Trade a 100 for
  10 tens. 14 tens take away 8 tens.
- Subtract the hundreds.
  6 hundreds take away 1 hundred.
- 8 thousands take away 2 thousands.

$$
\begin{array}{r}
{}^{6}8\,{}^{14}\cancel{7}\,{}^{10}\cancel{5}\,\cancel{0} \\
-\ 2\ 1\ 9\ 6 \\
\hline
6\ 5\ 6\ 4
\end{array}
$$

1 Solve the 4-digit subtractions without trading.

a
$$
\begin{array}{r}
8\ 6\ 4\ 9 \\
-\ 2\ 3\ 1\ 6 \\
\hline
\end{array}
$$

b
$$
\begin{array}{r}
7\ 7\ 4\ 8 \\
-\ 5\ 2\ 3\ 7 \\
\hline
\end{array}
$$

c
$$
\begin{array}{r}
8\ 6\ 8\ 4 \\
-\ \ \ \ 3\ 5\ 2 \\
\hline
\end{array}
$$

2 Complete these subtractions with trading in the ones.

a
$$
\begin{array}{r}
8\ 4\ 7\ 2 \\
-\ 3\ 2\ 3\ 4 \\
\hline
\end{array}
$$

b
$$
\begin{array}{r}
9\ 6\ 8\ 3 \\
-\ 4\ 5\ 6\ 6 \\
\hline
\end{array}
$$

c
$$
\begin{array}{r}
5\ 7\ 6\ 1 \\
-\ 3\ 3\ 3\ 5 \\
\hline
\end{array}
$$

d
$$
\begin{array}{r}
7\ 7\ 9\ 3 \\
-\ 2\ 4\ 7\ 4 \\
\hline
\end{array}
$$

e
$$
\begin{array}{r}
5\ 7\ 6\ 4 \\
-\ 2\ 5\ 3\ 6 \\
\hline
\end{array}
$$

3 These subtractions require trading in the ones or tens columns.

a
$$
\begin{array}{r}
6\ 7\ 4\ 3 \\
-\ 3\ 4\ 1\ 6 \\
\hline
\end{array}
$$

b
$$
\begin{array}{r}
8\ 6\ 4\ 2 \\
-\ 3\ 2\ 5\ 0 \\
\hline
\end{array}
$$

c
$$
\begin{array}{r}
8\ 5\ 2\ 0 \\
-\ 6\ 2\ 4\ 6 \\
\hline
\end{array}
$$

d
$$
\begin{array}{r}
7\ 7\ 3\ 2 \\
-\ 3\ 5\ 4\ 1 \\
\hline
\end{array}
$$

e
$$
\begin{array}{r}
5\ 6\ 4\ 2 \\
-\ 2\ 0\ 7\ 1 \\
\hline
\end{array}
$$

4

Populations and distances from Brisbane		
Town	**Population**	**From Brisbane**
Charleville	3400	744 km
Longreach	4152	1260 km
Normanton	1150	2152 km
Kuranda	750	1744 km
Mackay	58641	975 km

Calculate the difference in population between these centres.

a
Longreach	4152
Charleville	3400
Difference	

b
Charleville	
Normanton	
Difference	

c
Longreach	
Normanton	
Difference	

5 Which town is further from Brisbane? By how much?

a Longreach or Charleville?	**b** Normanton or Mackay?	**c** Mackay or Charleville?	**d** Normanton or Kuranda?
_____ km	_____ km	_____ km	_____ km

N3.2 Students identify and solve addition and subtraction problems involving whole numbers, and decimal fractions in context, selecting from a range of computation methods, strategies and known number facts.

25×3

To solve this I multiply the tens then the ones.

Hund	Tens	Ones
	2	5
×		3
	6	0
	1	5
	7	5

three 20s ⟶
three 5s ⟶
60 + 15 ⟶

6 Multiply the tens and then the ones to complete these multiplications.

a
Hund	Tens	Ones
	2	5
×		4

b
Hund	Tens	Ones
	2	7
×		4

c
Hund	Tens	Ones
	3	3
×		3

d
Hund	Tens	Ones
	4	4
×		5

e
Hund	Tens	Ones
	3	6
×		4

f
Hund	Tens	Ones
	5	4
×		4

g
Hund	Tens	Ones
	3	7
×		5

h
Hund	Tens	Ones
	4	5
×		6

i
Hund	Tens	Ones
	2	6
×		7

j
Hund	Tens	Ones
	1	9
×		8

k
Hund	Tens	Ones
	4	5
×		5

l
Hund	Tens	Ones
	6	3
×		6

m
Hund	Tens	Ones
	7	5
×		7

n
Hund	Tens	Ones
	3	8
×		8

o
Hund	Tens	Ones
	4	7
×		9

7 Calculate how much each worker would save.

a
Bank book
How much would Pedro save in 5 weeks if he saved $27 each week?

Hund	Tens	Ones
×		

$ _____

b
Bank book
Jean saved $36 each week for 7 weeks. What are her total savings?

Hund	Tens	Ones
×		

$ _____

N3.3 Students identify and solve multiplication and division problems involving whole numbers, and decimal fractions in context, selecting from a range of computation methods, strategies and known number facts.

99

Grid reference points

Grid reference points on a grid show where two lines meet.

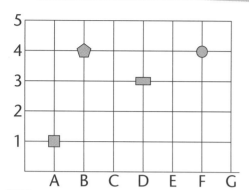

8 What shapes are found at these grid reference points?

a A1 _____

b D3 _____

c B4 _____

d F4 _____

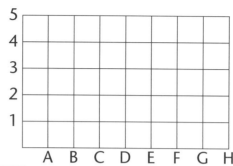

9 Join these points to form a shape.

Join B1 to F1
F1 to F4
F4 to B4
B4 to B1

Remember across before up.

Teachers' golf day

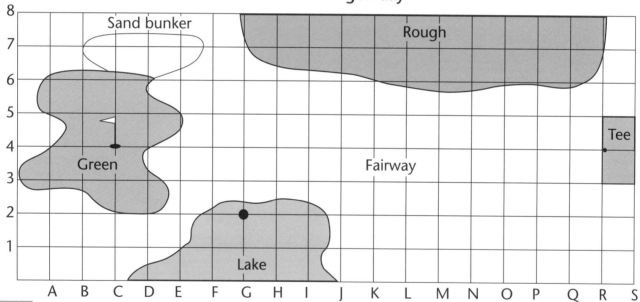

10 Use the grid reference points to draw the positions of the teachers' golf balls.

a Mrs Murray	G2 (done)	**e** Mr Irons	F4
b Ms Cooper	J8	**f** Ms Hill	Q2
c Mr Comerford	E7	**g** Mr Spears	A8
d Ms Norman	D5	**h** Ms Salter	L4

11 Tim's teacher, Mr Wilson, is a really awful golfer. He hit from the tee to N5, then into the rough at H7. From there he hit over the green to C1, then onto the green at B4. He then sunk his putt. Join up Mr Wilson's shots with a line to show how he played the hole.

S3.2 Students interpret and create maps and plans using a range of conventions, describe locations and give directions using major compass points, angles and grids.

12 Jordan has a bag of 20 lollies consisting of six different colours.

a Which colour is the most likely to be taken out of the party bag? _____

b Which colour is the least likely to be taken out of the party bag? _____

c Which colour is a little more likely to be taken out of the bag before green? _____

d Which two colours share the same likelihood of being taken out of the bag first? _____

e If all the blue lollies were eaten first which colour would then be the most likely to be taken out of the bag? _____

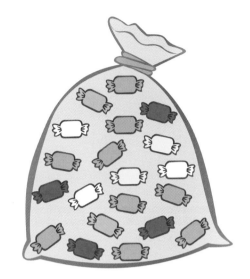

13 Colour the 15 lollies in this bag to match the tags at the side of the page.

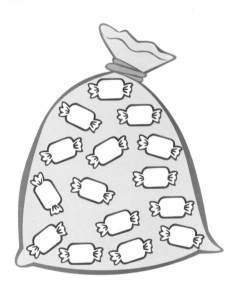

a Red has only one chance of being pulled out first.

b Blue has twice as many chances of being pulled out first compared to red.

c Green has four times more chances of being pulled out first compared to blue.

d Yellow has only half as many chances as green of being pulled out first.

14  Mike said, "There are 10 marbles in my bag. Five of them are red, one is blue and the others are green. It is equally likely that I will pick out a blue one first as it is I will pick out a red one."

Explain why you agree or disagree with Mike.

_____

_____

_____

_____

CD3.1 Students identify all possible outcomes of familiar situations or actions and, for these sample spaces, order the likelihood of occurrence of the identified outcomes using experimental data.

101

Extended multiplication

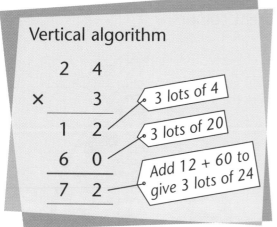

Vertical algorithm

```
    2  4
×      3        3 lots of 4
  ─────
    1  2        3 lots of 20
    6  0
  ─────        Add 12 + 60 to
    7  2        give 3 lots of 24
  ─────
```

1 Complete each multiplication by multiplying the ones and then the tens.

a
Hund	Tens	Ones
	2	3
×		3

b
Hund	Tens	Ones
	2	6
×		2

c
Hund	Tens	Ones
	3	4
×		2

d
Hund	Tens	Ones
	2	7
×		3

e
Hund	Tens	Ones
	2	8
×		4

f
Hund	Tens	Ones
	2	6
×		4

g
Hund	Tens	Ones
	2	8
×		5

h
Hund	Tens	Ones
	3	6
×		5

i
Hund	Tens	Ones
	2	7
×		6

j
Hund	Tens	Ones
	1	9
×		6

k
Hund	Tens	Ones
	2	5
×		6

l
Hund	Tens	Ones
	1	8
×		6

m
Hund	Tens	Ones
	2	3
×		6

2 Use some of the strategies opposite to complete the multiplications.

a 25 × 4 =

b 21 × 3 =

c 32 × 4 =

d 21 × 8 =

e 42 × 3 =

f 42 × 8 =

g 44 × 3 =

h 15 × 6 =

i 24 × 5 =

j 54 × 2 =

× 2	double
× 3	triple
× 4	double double
× 5	times 10 then half
× 6	times 3 then double
× 8	double double double

3 Solve these problems and check your solution with a calculator.

a Tatijana is saving $24 per week in order to buy a ring. How much money will she have after six weeks?

b Class 5G have baked 9 trays of cakes for the cake stall. How many cakes are there if 24 cakes are on each tray?

N3.3 Students identify and solve multiplication and division problems involving whole numbers, and decimal fractions in context, selecting from a range of computation methods, strategies and known number facts.

Patterns on a hundred chart

4 Follow the rule to extend the pattern.
Shade this pattern on the hundreds chart.

Rule: + 11

2	13	24					

5 Follow the rule to extend the pattern.
Shade this pattern on the hundreds chart.

Rule: × 11

1	2	3	4	5	6	7	8	9
11	22							

6 Why would it be easy to spot a mistake in the 11's facts on the chart?

_____

_____

1	2	3	4	5	6	7	8	9	10
11	12	13	14	15	16	17	18	19	20
21	22	23	24	25	26	27	28	29	30
31	32	33	34	35	36	37	38	39	40
41	42	43	44	45	46	47	48	49	50
51	52	53	54	55	56	57	58	59	60
61	62	63	64	65	66	67	68	69	70
71	72	73	74	75	76	77	78	79	80
81	82	83	84	85	86	87	88	89	90
91	92	93	94	95	96	97	98	99	100

7 Explain the shaded pattern on the chart opposite.

_____

_____

1	2	3	4	5	6	7	8	9	10
11	12	13	14	15	16	17	18	19	20
21	22	23	24	25	26	27	28	29	30
31	32	33	34	35	36	37	38	39	40
41	42	43	44	45	46	47	48	49	50
51	52	53	54	55	56	57	58	59	60
61	62	63	64	65	66	67	68	69	70
71	72	73	74	75	76	77	78	79	80
81	82	83	84	85	86	87	88	89	90

8 Circle this pattern on the same chart.
Start at 5, then add 3, then continue to add 5 and then 3 until you run out of space.

9 Sam was experimenting with patterns and created this pattern.
Continue the third line.

10	20	30	40	50	60	70	80	90
−1	−2	−3	−4	−5	−6	−7	−8	−9
9	18	27						

I think I've created a set of multiplication facts.

10 Explain the pattern that Sam created.

_____

_____

PA3.1 Students create and continue number patterns, identify, describe and represent relationships between two quantities and use backtracking to reverse any one of the four operations.

103

11 Peter said that most children are born in the month of September.

a Make a prediction about the month in which most students in your class were born. _____

b Make a prediction about the month in which the least students were born. _____

c Conduct a survey of the students in your class to find out if Peter was correct. Record your information using tally marks on the grid.

January	February	March	April	May	June

July	August	September	October	November	December

d Construct a bar graph to display the data.

I was born in February.

Months in which students in our class were born

Number of students (y-axis 0–11) vs Months (Jan, Feb, March, April, May, June, July, August, Sept, October, Nov, Dec)

12 Was Peter correct about September being the month in which most children are born? _____

13 Were the survey results as you expected them to be? _____

14 Write a comment about your survey results. _____

CD3.2 Students design and trial a variety of data collection methods and use existing sources of data to investigate their own and others' questions, organise data and create suitable displays identifying and interpreting elements of the displays.

Litres and millilitres

15 Three sets of beakers have been filled with red, green and orange cordial. Record in millilitres, then litres and millilitres, how much cordial is in each pair.

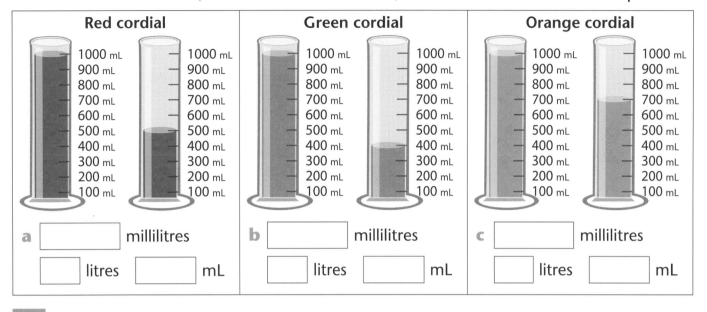

Red cordial	Green cordial	Orange cordial

a [_____] millilitres

[____] litres [____] mL

b [_____] millilitres

[____] litres [____] mL

c [_____] millilitres

[____] litres [____] mL

16 How many more millilitres are needed to make 1 litre?

200 mL   250 mL   350 mL   500 mL   280 mL   600 mL   240 mL   180 mL

	Container	Millilitres
a	Cream	
b	Shampoo	
c	Jam	
d	Milk	

	Container	Millilitres
e	Can	
f	Sugar soap	
g	Glass	
h	Cup	

17 Convert the millilitre measurements to litres and millilitres. The first one has been done for you.

1500 mL is the same as 1.5 L.

	Millilitres	Litres
a	1 800 mL	1.8 L
b	1 600 mL	
c	1 900 mL	
d	1 250 mL	

	Millilitres	Litres
e	2 300 mL	
f	5 400 mL	
g	3 450 mL	
h	4 500 mL	

M3.1 Students identify and use equivalent forms of standard units when measuring, comparing and ordering, and estimate using a range of personal referents.

105

Contracted multiplication

Hund	Tens	Ones
	2	
	5	3
×		9
4	7	7

9 × 3 = 27. Write the 7 in the ones column and trade the 2 to the tens column.

9 × 5 tens = 45 plus the 2 tens traded = 47 tens. Write a 7 in the tens column and a 4 in the hundreds column.

1 Complete the multiplications. The first one is done for you.

a

Hund	Tens	Ones
	4	
	2	8
×		5
1	4	0

b

Hund	Tens	Ones
	2	5
×		3

c

Hund	Tens	Ones
	4	3
×		5

d

Hund	Tens	Ones
	3	7
×		8

e

Hund	Tens	Ones
	2	9
×		4

f

Hund	Tens	Ones
	2	4
×		9

g

Hund	Tens	Ones
	4	3
×		8

h

Hund	Tens	Ones
	8	2
×		7

i

Hund	Tens	Ones
	6	3
×		6

j

Hund	Tens	Ones
	3	4
×		5

k

Hund	Tens	Ones
	2	9
×		8

l

Hund	Tens	Ones
	4	8
×		6

m

Hund	Tens	Ones
	7	2
×		3

n

Hund	Tens	Ones
	5	1
×		4

o

Hund	Tens	Ones
	6	3
×		7

2 Solve these problems using any strategy you wish.

a	How many children were at the carnival if there were 8 schools with 54 competitors each?	
b	During the 9 games of the netball tournament Mandy averaged 28 points per game. What was her total number of points?	
c	How much has David saved if he has made 8 deposits of $78 into his bank account?	

d	What is the distance around the outside of the square garden bed if the sides are 23 metres long?	
e	How far does Violet swim per fortnight if she trains 6 days a week and completes 4 kilometres each session?	
f	How many metres did Ebony swim if she completed 8 laps of the 49-metre pool?	

N3.3 Students identify and solve multiplication and division problems involving whole numbers, and decimal fractions in context, selecting from a range of computation methods, strategies and known number facts.

EFTPOS stands for Electronic Funds Transfer at Point of Sale.
This means that people don't need to pay for things with cash but can swipe their bank card through an electronic machine which transfers the money from the customer's account to the bank account of the business.

> I can swipe my card and $49.99 will be withdrawn from my bank account, allowing me to buy a new jacket costing $49.99.

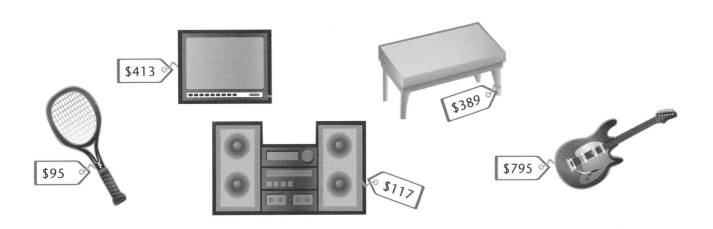

$413

$389

$95

$117

$795

3 Calculate how much money has been transferred from the customer's account in each of these transactions. In some cases the customer has also taken out extra cash.

	Transaction	EFTPOS amount
a	Bought a tennis racquet and took out $100 cash.	
b	Bought a computer and took out $200 cash.	
c	Bought a CD player and book out $50 cash.	
d	Bought a coffee table and took out $20 cash.	
e	Bought the guitar and took out $200 cash.	
f	Bought two computers.	
g	Bought four tennis racquets.	

> $95 + $100
> That's …

4 What are the advantages of using EFTPOS cards when shopping? _____

_____

5 Could there be any disadvantages with electronic banking? _____

_____

N3.2 Students identify and solve addition and subtraction problems involving whole numbers, and decimal fractions in context, selecting from a range of computation methods, strategies and known number facts.

107

6 Use the square corner of a piece of paper, which is a right angle, as an angle tester to classify the angles as acute, straight, obtuse or reflex. Reflex angles are between 180° and 360°.

Angle tester

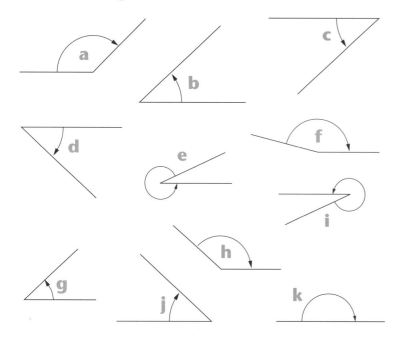

	Acute	Straight	Obtuse	Reflex
a				
b				
c				
d				
e				
f				
g				
h				
i				
j				
k				

7 Circle all acute angles, put a cross on all obtuse angles and put an 'R' on all right angles on the map.

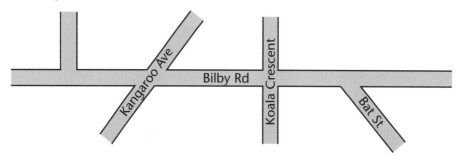

8 Look carefully at the two angles in the box.

a Are the two angles the same size? _____

b Explain why. _____

_____

_____

_____

_____

_____

_____

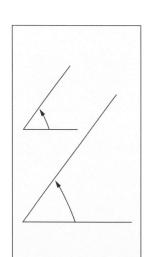

I found this one the same.

S3.1 Students describe the defining geometric properties of families of 3D shapes, model 3D shapes using nets and other representations, and identify and describe the properties of specific families and subgroups of 2D shapes.

9 Study this calendar in order to name the day for the dates below.

January

Sun		2	9	16	23	30
Mon		3	10	17	24	31
Tues		4	11	18	25	
Wed		5	12	19	26	
Thur		6	13	20	27	
Fri		7	14	21	28	
Sat	1	8	15	22	29	

February

Sun			6	13	20	27
Mon			7	14	21	28
Tues		1	8	15	22	
Wed		2	9	16	23	
Thur		3	10	17	24	
Fri		4	11	18	25	
Sat		5	12	19	26	

a 19 January _____

b 13 January _____

c 14 February _____

d 26 January _____

e 28 February _____

f 18 February _____

g 29 January _____

h 1 February _____

This calendar has the days of the week going across the top.

January

S	M	T	W	T	F	S
						1
2	3	4	5	6	7	8
9	10	11	12	13	14	15
16	17	18	19	20	21	22
23	24	25	26	27	28	29
30	31					

February

S	M	T	W	T	F	S
		1	2	3	4	5
6	7	8	9	10	11	12
13	14	15	16	17	18	19
20	21	22	23	24	25	26
27	28					

March

S	M	T	W	T	F	S
		1	2	3	4	5
6	7	8	9	10	11	12
13	14	15	16	17	18	19
20	21	22	23	24	25	26
27	28	29	30	31		

10 Calculate the days from:

a Sunday 2 January to Sunday 16 January. _____

b Monday 10 January to Thursday 20 January. _____

c Friday 7 January to Wednesday 26 January. _____

d Friday 21 January to Friday 4 February. _____

e The first Tuesday in February to Thursday 17 February. _____

Dates are often represented in this form: **dd/mm/yy**. For example, Tom was born on the 23rd of June 1998. When recorded in digital form it becomes 23/06/98.

11 Use the digital form to express the date of birth of the people in this family.

	Name	Date of birth	Digital form
a	Brett Martini	Fifth of September, 1972	
b	Susan Martini	Thirtieth of October, 1976	
c	Jude Martini	Nineteenth of November, 1997	
d	Leah Martini	Eleventh of January, 2001	

12 Write today's date using the dd/mm/yy form. [ ]

M3.2 Students read, record and calculate with 12-hour time, and interpret calendars and simple timetables related to daily activities.

109

Division strategies

Sometimes divisions don't work out equally and have **remainders**.
Let's see how Mrs Flockhart shared 73 cakes among 3 classes.

73 shared among 3. 3⟌73̄	Share out the tens with each class getting 2. 　　2 3⟌7¹3	Trade the 1 ten left for 10 ones. Now share the 13 ones among 3. 　　2 4 r 1 3⟌7¹3 Answer: 24 remainder 1

1 Solve the divisions.

a　　　　r
　3⟌43̄

b　　　　r
　4⟌57̄

c　　　　r
　5⟌66̄

d　　　　r
　4⟌54̄

e　　　　r
　3⟌74̄

f　　　　r
　5⟌67̄

g　　　　r
　4⟌61̄

h　　　　r
　6⟌81̄

i　　　　r
　6⟌85̄

j　　　　r
　5⟌87̄

k　　　　r
　6⟌92̄

l　　　　r
　7⟌86̄

m　　　　r
　8⟌97̄

n　　　　r
　5⟌82̄

o　　　　r
　7⟌99̄

2 Use the sharing strategy outlined opposite to try to solve these mentally.

a $45 \div 3 =$ _____

b $52 \div 4 =$ _____

c $96 \div 4 =$ _____

d $72 \div 6 =$ _____

e $84 \div 7 =$ _____

f $91 \div 7 =$ _____

g $78 \div 6 =$ _____

h $42 \div 3 =$ _____

i $56 \div 4 =$ _____

j $65 \div 5 =$ _____

k $64 \div 4 =$ _____

l $70 \div 5 =$ _____

m $78 \div 6 =$ _____

n $98 \div 7 =$ _____

o $96 \div 8 =$ _____

Share 65 among 5.
• 6 tens ÷ 5 = **1 ten** and then 1 ten left over.
• 1 ten and 5 ones divided by 5 = **3 ones**.
Answer = 13.

3 Use the halving strategy outlined opposite to divide these larger numbers.

a $160 \div 2 =$ _____

b $160 \div 4 =$ _____

c $160 \div 8 =$ _____

d $240 \div 2 =$ _____

e $240 \div 4 =$ _____

f $240 \div 8 =$ _____

g $200 \div 2 =$ _____

h $200 \div 4 =$ _____

i $200 \div 8 =$ _____

j $224 \div 2 =$ _____

k $224 \div 4 =$ _____

l $224 \div 8 =$ _____

÷2 halve,
÷4 halve and halve again,
÷8 halve, halve and halve again.

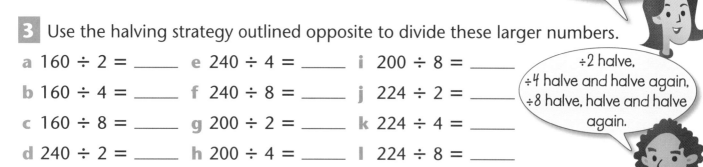

4 Make a list of all the ways that 120 counters can be evenly shared.

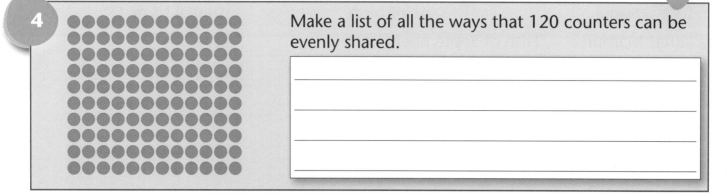

N3.3 Students identify and solve multiplication and division problems involving whole numbers, and decimal fractions in context, selecting from a range of computation methods, strategies and known number facts.

5 Select the ⬜= sign or the ⬜≠ sign to describe these sentences.

a 13 + 27 − 9	⬜	65 − 20 − 8	
b 26 + 48 − 7	⬜	80 + 15 − 30	
c 64 − 10 + 8	⬜	52 + 20 − 8	
d 99 − 66 + 33	⬜	30 + 30 + 6	

e 5 × 3 × 2	⬜	4 × 5 × 2	
f 4 × 7 × 5	⬜	7 × 2 × 10	
g 100 ÷ 4 × 2	⬜	5 × 10 × 2	
h 64 × 1 ÷ 8	⬜	3 × 3 × 3	

6 Use the ⬜= sign or the ⬜≠ sign to describe these shopping situations.

a In one pan I have three 80-gram apricots. In the other pan I have four 60-gram apricots.	⬜
b In one pan there are three 100-gram chocolates. In the other pan there are ten 30-gram chocolates.	⬜
c In one pan there are five 80-gram peaches. In the other pan there are seven 60-gram peaches.	⬜

7 Use the greater than ⬜> , less than ⬜< or equals sign to compare these number sentences.

a 5 × 3 × 2 ⬜ 2 × 6 × 2   **f** 48 + 9 + 12 ⬜ 5 × 7 × 2

b 3 × 6 × 5 ⬜ 100 − 18   **g** 63 − 48 − 10 ⬜ 100 ÷ 2 ÷ 2

c 4 × 7 × 3 ⬜ 100 − 84   **h** 50 + 40 + 30 ⬜ 10 × 2 × 6

d 5 × 9 × 2 ⬜ 2 × 5 × 10   **i** 85 − 40 + 15 ⬜ 10 × 3 × 5

e 3 × 2 × 7 ⬜ 2 × 8 × 2   **j** 75 − 30 + 22 ⬜ 17 × 2 × 2

69 70
48 + 9 + 12 < 5 × 7 × 2

8 Grace won a $2 000 gift voucher to be spent at a department store. She is planning how to spend the money and has come up with several combinations.

Use the ⬜> and ⬜< symbols to complete these number sentences.

a $1 000 + $600 + $500 ⬜ $2 000

b $800 + $800 + $350 ⬜ $2 000

c $500 × 5 ⬜ $2 000

d $200 × 7 + $500 ⬜ $2 000

Gift Voucher
for $2000.00
Fountaingate

Letters for naming shapes, intervals and lines

Part of a line between two points is called an **interval**. Capital letters are used to name intervals.

Example: A •———————————————• B Interval AB is 50 millimetres.

9 Measure and label these intervals using capital letters. (When lettering a set of intervals, begin with A and B and continue through the alphabet.)

a •———————————————•

Interval ▢ = _____ mm

b •———————————————•

Interval ▢ = _____ mm

c •———————•

Interval ▢ = _____ mm

d •———————————————•

Interval ▢ = _____ mm

A line is named by placing letters on any two points on the line. Arrowheads are drawn at each end to show that lines are continuous.

Example: ←———C———————D———→ Line CD

10 Use capital letters to name points on these lines. Add arrowheads to show that lines are continuous. Name the lines.

a _____

Line ▢

b _____

Line ▢

c _____

Line ▢

Polygons are named by lettering the vertices.

Example: A ————————— B Triangle ABC
with C at top

11 Label each vertex and then use those letters to name the polygon. The first one has been done for you.

a D———————C

A———————B

Parallelogram ABCD

b

_____

c

_____

d

_____

e

_____

f

_____

S3.1 Students describe the defining geometric properties of families of 3D shapes, model 3D shapes using nets and other representations, and identify and describe the properties of specific families and subgroups of 2D shapes.

12 Peter has 3 marbles in the bag that are coloured red, yellow and blue. There are 6 different ways that the marbles can be drawn out of the bag. Colour the marbles to show the possible combinations that he could draw out.

Hint: (R)(Y)(B) is different from (B)(Y)(R)

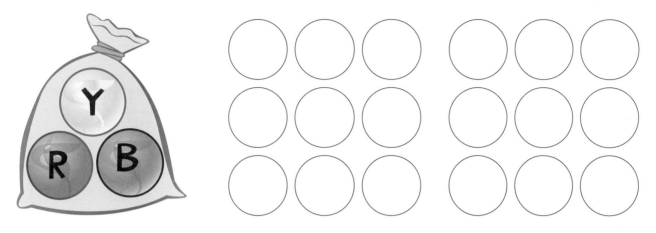

13 Green Bay netball club is changing its uniform and has a choice of 3 different coloured tops and 3 different coloured skirts.

Colour the uniforms to show all the different colour combinations it could have.

14 The nets of four cubes have been made for a colour game. Colour the faces on the nets to match the chance of that cube showing red.

red

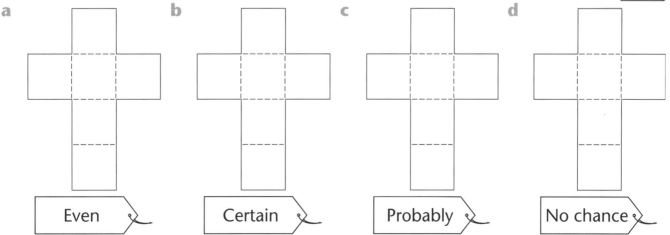

a   b   c   d

Even   Certain   Probably   No chance

Diagnostic review 3

PART 1

Complete the additions and subtractions.

a
```
  2 5 3 4
+ 3 3 4 7
_____
```

b
```
  5 2 8 7
+ 2 6 5 1
_____
```

c
```
  8 9 4 9
- 4 1 6 0
_____
```

d
```
  6 8 5 5
-   4 6 2
_____
```

e
```
  4 3 6 0
- 2 4 7 4
_____
```

f
```
  5 4 7 1
- 1 8 3 5
_____
```

g
```
  7 3 5 6
    2 6 2
+     7 8
_____
```

h
```
  2 5 4 8
      7 3
+ 3 5 9 3
_____
```

i
```
  2 0 7 6
  3 5 6 4
+ 2 5 3 9
_____
```

PART 2

Complete the multiplications and divisions.

a
```
    2 6
×    3
_____
```

b
```
    3 8
×    5
_____
```

c
```
    4 7
×    4
_____
```

d 26 ÷ 5 = ☐   remainder ☐

e 32 ÷ 6 = ☐   remainder ☐

f 41 ÷ 5 = ☐   remainder ☐

g 38 ÷ 4 = ☐   remainder ☐

h 3)24

i 4)32

j 5)35

k 6)26

l 7)24

m 8)35

n Cook House has 14 points. Increase their points $1\frac{1}{2}$ times. ☐

PART 3

Complete the number patterns.

Rule: ÷ 2

a

☐	80	70	60	50	40	30
▲	40	35				

Rule: × 3

b

▲	10	20	30	40	50	60
☐						

PART 4

Order the decimals from smallest to largest.

a	0.3	0.7	0.2	
b	1.2	1.5	1.3	
c	0.27	0.35	0.47	
d	1.27	1.3	1.2	

Round the decimals to the nearest whole number.

e 4.92 = ____   f 5.01 = ____

g 3.13 = ____   h 0.99 = ____

i $5.95 = ____   j $8.50 = ____

Describe each grid using tenths, hundredths and decimal fractions.

k  $\frac{}{10}$ $\frac{}{100}$ ☐

l 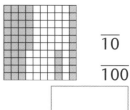 $\frac{}{10}$ $\frac{}{100}$ ☐

Use <, = or > to answer the following.

m $\frac{4}{10}$ ☐ $\frac{50}{100}$ n 0.60 ☐ $\frac{6}{10}$

o $\frac{80}{100}$ ☐ 0.90   p $\frac{1}{2}$ ☐ 0.50

Diagnostic review 3

PART 5
M3.1

Add another arm to each geo-strip to create an angle that matches the labels.

a   b   c

acute   right   obtuse

PART 6
S3.1

Sketch each shape after it has moved.

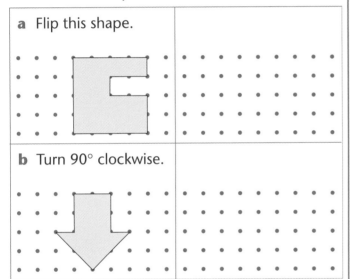

a Flip this shape.

b Turn 90° clockwise.

PART 7
S3.1

How can you tell the difference between an isosceles triangle and an equilateral triangle?

_____

_____

_____

PART 8
S3.2

a Put a cross on C5.

b Draw a circle on B3.

c Colour the square inside these grid references.

A5, B5, A4 and B4

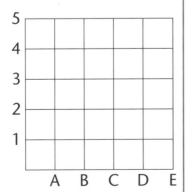

PART 9
M3.2

How many minutes to the next hour?

a [1][0]:[5][0] _____ minutes

b [1][1]:[4][0] _____ minutes

c [   ][9]:[3][0] _____ minutes

d [   ][7]:[2][0] _____ minutes

e [   ][5]:[2][5] _____ minutes

PART 10
MS3.1

a Colour the 1-litre container to the correct level if all the containers below were poured into it.

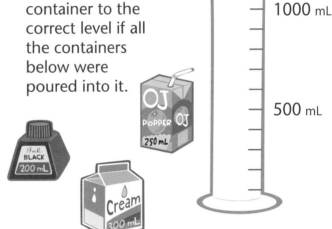

1000 mL

500 mL

How many millilitres in:

b	1 litre and 500 mL	mL
c	2 litres and 375 mL	mL

PART 11
CD3.1

True or false

When 2 dice are tossed:

a 7 has more chances than all of being tossed. _____

b Odd numbers and even numbers have the same chance of being tossed. _____

c 2 and 12 have the least chance of being tossed. _____

Rounding to the nearest 5 cents

In Australia we don't have 1 cent and 2 cent coins. This means that prices are rounded to the nearest 5 cents.
Prices ending in 3 cents are rounded to 5 cents and those ending in 8 cents are rounded to 10 cents.

1 Here are the prices of flavoured milk being sold at various shops. Below each carton write the price the customer would be charged after the price was rounded to the nearest 5 cents.

a $2.01   b $2.02   c $2.03   d $2.04   e $2.06   f $2.07   g $2.08   h $2.09

$_____ $_____ $_____ $_____ $_____ $_____ $_____ $_____

i What price was the most common price for the flavoured milk? _____

j What is the difference between the highest price and the lowest price after rounding? _____

If a customer buys several items at a shop it is only the total that is rounded.

 $1.23 $0.28 $0.48    $0.32 $1.37 $10.92

2 Calculate the total price of each shopping docket and then round it to the nearest 5 cents.

a
Comic book	$1.23
Pencil	$0.28
Maths book	$10.92
Pen	$1.37
Ruler	$0.48
Total	$
Rounded	$

b
Pen	$1.37
Ruler	$0.48
Ruler	$0.48
Ruler	$0.48
Comic book	$1.23
Total	$
Rounded	$

c
Maths book	$10.92
Comic book	$1.23
Pen	$1.37
Pen	$1.37
Ruler	$0.48
Total	$
Rounded	$

d
Maths book	$10.92
Eraser	$0.32
Comic book	$1.23
Comic book	$1.23
Comic book	$1.23
Total	$
Rounded	$

3 Calculate the amount of change after rounding the total cost.

a How much change from $10 would I get if I bought a comic book and a pencil? _____

b How much change from $5 would I get if I bought 3 erasers and a pencil? _____

4 Examine some shopping dockets at home to see whether rounding occurs when electronic cards are used. What did you notice?

_____

Adding and subtracting decimals

Remember to keep the decimal points underneath each other when adding or subtracting decimals.

5 Complete the additions and subtractions.

a	b	c	d	e
5 . 6 4	3 . 5 7	4 . 3 3	2 5 . 4 4	4 4 . 7 7
+ 3 . 3 2	+ 6 . 3 8	+ 5 . 2 8	+ 6 4 . 8 3	+ 5 2 . 8 7

f	g	h	i	j
3 9 . 7 8	5 8 . 5 4	8 7 . 8 4	8 6 . 6 5	6 7 . 6 3
− 1 4 . 4 3	− 1 3 . 3 5	− 2 4 . 6 6	− 4 2 . 7	− 3 4 . 8 6

k	l	m	n	o
2 3 . 2 1	2 1 . 0 4	1 3 . 2 6	2 3 . 3 1	2 3 . 1 7
1 4 . 3 2	3 3 . 2 2	3 2 . 2 7	7 0 . 6 8	3 2 . 6 7
+ 3 1 . 4 4	+ 1 2 . 5 8	+ 2 2 . 4 5	+ 5 . 7 4	+ 4 1 . 2 7

lunch seat 3.735 m 　　table 1.364 m 　　lounge 2.264 m 　　car 4.650 m

6 Use the measurements to solve the problems. The first one is started for you.

a What is the total length of the lunch seat and the lounge if placed end to end?	3 . 7 3 5 + 2 . 2 6 4	**b** What is the total length of the table and car if placed end to end?	.
c What is the difference between the length of the car and the length of the table?	.	**d** What is the difference between the length of the lunch seat and the length of the lounge?	.

7 Solve the problems.

a Jess is 1.738 m tall and Kevin is 1.382 m tall. How much taller is Jess than Kevin?		**b** The potatoes weighed 28.46 kg and the onions weighed 14.35 kg. How much heavier are the potatoes?	

Compass points

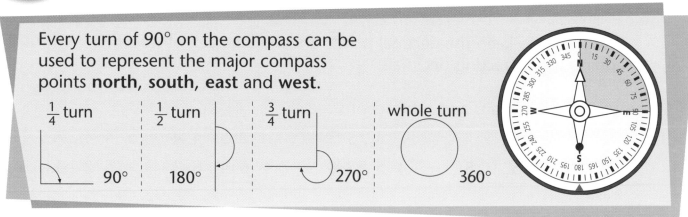

Every turn of 90° on the compass can be used to represent the major compass points **north**, **south**, **east** and **west**.

$\frac{1}{4}$ turn	$\frac{1}{2}$ turn	$\frac{3}{4}$ turn	whole turn
90°	180°	270°	360°

8 Use your knowledge of quarter turns, half turns, three-quarter turns and full turns to answer these questions about compass points. How many degrees are there between:

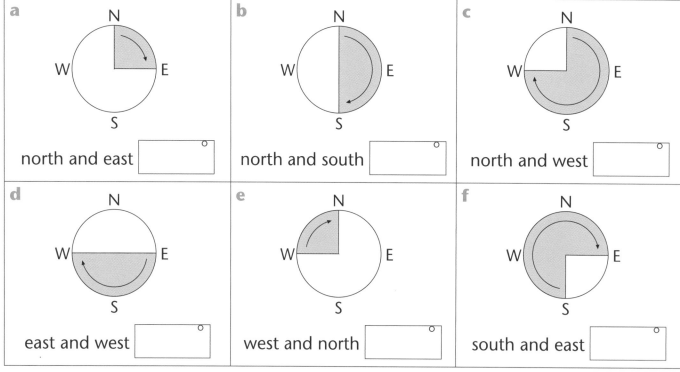

a

north and east [ ]°

b

north and south [ ]°

c

north and west [ ]°

d

east and west [ ]°

e

west and north [ ]°

f

south and east [ ]°

9

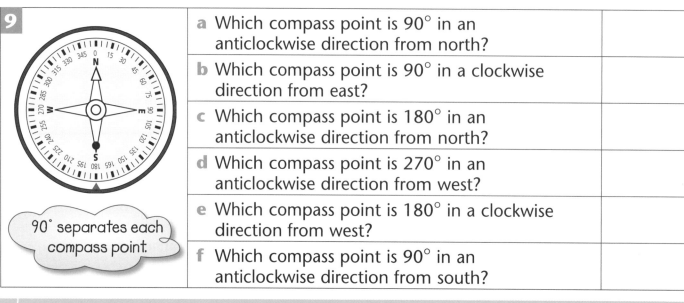

90° separates each compass point.

a Which compass point is 90° in an anticlockwise direction from north?	
b Which compass point is 90° in a clockwise direction from east?	
c Which compass point is 180° in an anticlockwise direction from north?	
d Which compass point is 270° in an anticlockwise direction from west?	
e Which compass point is 180° in a clockwise direction from west?	
f Which compass point is 90° in an anticlockwise direction from south?	

S3.2 Students interpret and create maps and plans using a range of conventions, describe locations and give directions using major compass points, angles and grids.

10 Apply the rule to complete the output column on these function machines.

a

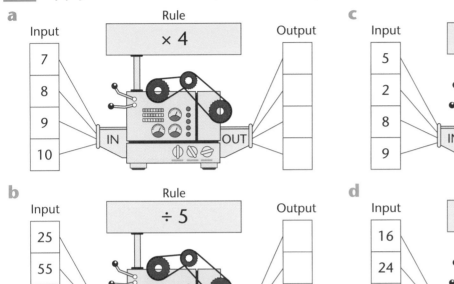

Input: 7, 8, 9, 10
Rule: × 4
Output

c

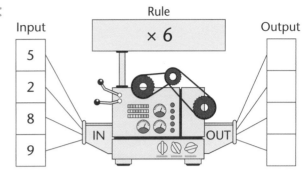

Input: 5, 2, 8, 9
Rule: × 6
Output

b

Input: 25, 55, 45, 35
Rule: ÷ 5
Output

d

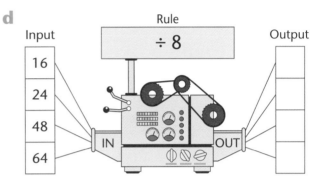

Input: 16, 24, 48, 64
Rule: ÷ 8
Output

11 Reverse the rule on each function machine and apply it to the output numbers to work out what number belongs in the input box.

a

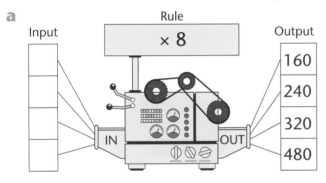

Input
Rule: × 8
Output: 160, 240, 320, 480

b

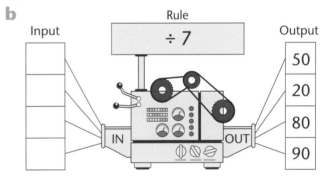

Input
Rule: ÷ 7
Output: 50, 20, 80, 90

12 Celebrity Heads

Use the clues to work the celebrity numbers.

a

If you add 8 my total would be 20.

b

If you divide me by 5 my quotient would be 17.

c

If you multiplied me by 18 my product would be 90.

d

If you multiplied me by myself my product would be 81.

Multiplying tens

Hund	Tens	Ones
	4	0
×		5
2	0	0

4 tens × 5 = 20 tens or 200

1 Multiply these multiples of ten.

a
Hund	Tens	Ones
	3	0
×		5

b
Hund	Tens	Ones
	5	0
×		4

c
Hund	Tens	Ones
	6	0
×		6

d
Hund	Tens	Ones
	7	0
×		2

e
Hund	Tens	Ones
	4	0
×		3

f
Hund	Tens	Ones
	7	0
×		5

g
Hund	Tens	Ones
	6	0
×		8

h
Hund	Tens	Ones
	2	0
×		9

i
Hund	Tens	Ones
	5	0
×		3

j
Hund	Tens	Ones
	5	0
×		5

k
Hund	Tens	Ones
	7	0
×		8

l
Hund	Tens	Ones
	9	0
×		2

m
Hund	Tens	Ones
	5	0
×		7

n
Hund	Tens	Ones
	6	0
×		4

o
Hund	Tens	Ones
	5	0
×		6

p
Hund	Tens	Ones
	7	0
×		4

q
Hund	Tens	Ones
	8	0
×		6

r
Hund	Tens	Ones
	8	0
×		8

s
Hund	Tens	Ones
	6	0
×		7

t
Hund	Tens	Ones
	8	0
×		5

Sleepers
$20

Plants
$30

Soil per tonne
$40

Paving per square metre
$50

2 A landscape gardener ordered the following materials. How much did they cost him?

a 7 sleepers _____

b 6 plants _____

c 8 tonnes of soil _____

d 9 square metres of paving _____

N3.3 Students identify and solve multiplication and division problems involving whole numbers, and decimal fractions in context, selecting from a range of computation methods, strategies and known number facts.

Rounding to 1 000

In Unit 13 we learned how to estimate by rounding numbers to 100. We can now estimate larger numbers by rounding them to the nearest 1 000.

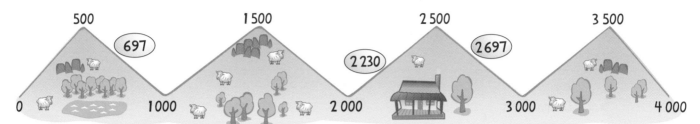

3 Write the closest 1 000 that each number will slide to.

a (697) _____   b (2 230) _____   c (2 697) _____

4 Round each number to the nearest 1 000. Numbers ending in 500 are rounded up.

a 1 979 _____   d   679 _____   g 5 683 _____   j 3 597 _____

b   876 _____   e   356 _____   h 7 500 _____   k 4 500 _____

c 1 230 _____   f 4 777 _____   i 3 500 _____   l 7 499 _____

5 We should always check our answers by estimating to see if they are correct. Estimate by rounding to 1 000 to check if these answers are reasonable.

	Question	Answer	Estimate	Reasonable or unreasonable
a	3 799 + 2 199 =	6 090	6 000	reasonable
b	2 074 + 2 963 =	5 037		
c	4 519 + 3 448 =	7 967		
d	7 776 + 2 335 =	12 111		
e	8 894 + 4 077 =	12 971		
f	4 896 + 3 075 =	6 091		

2 999 + 4 041 That's easy! 3 000 + 4 000 = 7 000

6 Round each price to the nearest dollar to estimate the total price.

a One tin of soup at $5.75 and a jar of coffee at $4.35 $ _____

b One pack of dog food at $9.95 and a pack of biscuits at $3.10 $ _____

c One pair of shorts at $29.65 and a top at $14.40? $ _____

d One main meal at $16.75 and a dessert at $9.30 $ _____

e One TV at $694.99 and a DVD player at $299.99 $ _____

f Five packs of dog food at $9.95 and six tins of soup at $5.75 $ _____

Symbols on picture graphs can be used to show more than 1 item.

7 Use the key to answer questions on the picture graph.

Transport to school

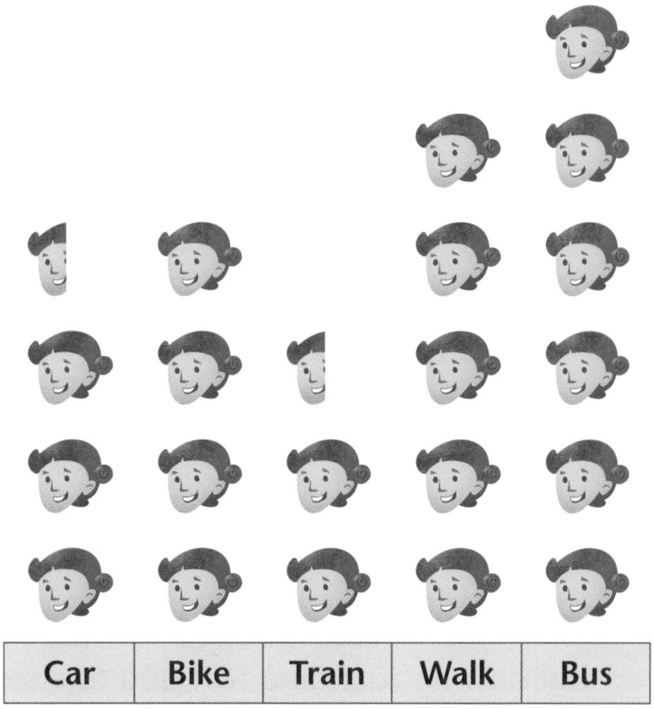

Key
= 10 children

a How many students rode a bike to school? _____

b How many students caught the bus to school? _____

c How many students caught the train to school? _____

d How many students walked to school? _____

e How many students were driven in a car? _____

f How many students are displayed on the graph? _____

Car	Bike	Train	Walk	Bus

8 Describe some conditions that might affect this data.

_____

_____

9 Carry out your own transport investigation. First decide how you are going to gather your data and which group you are going to target. Present your data in a table and as a picture graph.

My target group was: _____

Type of transport	Tally

CD3.2 Students design and trial a variety of data collection methods and use existing sources of data to investigate their own and others' questions, organise data and create suitable displays identifying and interpreting elements of the displays.

10 Billy and Zena drew this large picture with chalk on the playground. They wanted to know how big it was, so they drew square metres in chalk over the drawing.

a How many square metres did the drawing cover? _____

b Can you see a quick way to count the square metres? _____

c Explain how you counted the squares.

_____

_____

_____

11 Construct a handball court on the school playground.

What you will need	Dimensions of court
Metre rule Chalk A large 90° angle tester A small group of children to help you A day when it isn't raining	6 m wide, 4 m tall, 2 m, 3 m

12 With a partner, work out a way to calculate these areas of your handball court.

a The area of the whole court _____

b The area of half the court _____

c The area of a quarter of the court _____

13 Check the accuracy of your court by measuring its diagonals. They should be the same.

M3.1 Students identify and use equivalent forms of standard units when measuring, comparing and ordering, and estimate using a range of personal referents.

123

Division

1 Solve the divisions.

a	e	i	m	q
3)21	4)23	3)48	7)84	4)61

b	f	j	n	r
4)28	5)34	4)60	8)99	5)72

c	g	k	o	s
5)40	6)26	5)60	5)68	6)80

d	h	l	p	t
6)30	7)37	5)90	6)69	7)93

YUM YUM
APPLES Pty Ltd

YUM YUM APPLES 48 APPLES

48 APPLES

2 Con the fruit shop owner wants to know how many bags of different quantities he could make up from the box of 48 apples he bought at the market.

a How many bags of 3 could he make? _____

b How many bags of 4 could he make? _____

c How many bags of 5 could he make? _____

d How many bags of 6 could he make? _____

e How many bags of 8 could he make? _____

f How many bags of 9 could he make? _____

3 Show 4 ways that the jar of 37 lollies could be shared.

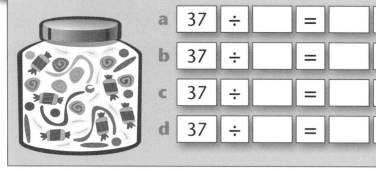

a 37 ÷ [ ] = [ ] r [ ]

b 37 ÷ [ ] = [ ] r [ ]

c 37 ÷ [ ] = [ ] r [ ]

d 37 ÷ [ ] = [ ] r [ ]

Knowing your multiplication facts makes division easy.

4 Solve the quick division problems.

a 32 stickers shared among 4 children _____

b 37 chips shared among 5 children _____

c 43 chocolates shared among 8 boys _____

d 65 bananas shared among 4 girls _____

N3.3 Students identify and solve multiplication and division problems involving whole numbers, and decimal fractions in context, selecting from a range of computation methods, strategies and known number facts.

5 Complete these subtractions.

a
```
  9 6 8 5
- 3 2 7 1
─────────
```

b
```
  8 5 4 7
- 3 2 2 1
─────────
```

c
```
  8 6 5 8
- 2 5 4 3
─────────
```

d
```
  7 7 6 7
- 5 4 2 4
─────────
```

e
```
  5 4 6 7
- 3 2 5 4
─────────
```

6 These subtractions require trading in the ones or tens columns.

a
```
  4 4 8 3
- 2 1 6 8
─────────
```

b
```
  5 6 4 8
- 3 4 9 6
─────────
```

c
```
  9 7 6 4
- 2 3 4 9
─────────
```

d
```
  8 8 5 6
- 7 1 8 3
─────────
```

e
```
  6 4 6 4
- 3 1 3 8
─────────
```

7 Use trading in the tens and hundreds columns to complete these.

a
```
  9 3 7 8
- 4 8 3 2
─────────
```

b
```
  8 2 8 6
- 5 9 5 4
─────────
```

c
```
  9 6 9 4
- 3 9 2 9
─────────
```

d
```
  8 9 3 9
- 5 8 6 8
─────────
```

e
```
  7 6 5 4
- 4 9 2 7
─────────
```

HOT DEALS
ON WHEELS this week at **MOTOWN MOTORS**

BARINA $9860
Excellent condition. New tyres. Bright orange.

BMW $5905
1998 model
Needs repairs. Rust colour

COLT $8905
2002 model
Bright red. Sun roof.

TOYOTA $4390
2000 model. 30 000 km. Looks good. Runs well.

FORD $6190
2003 wagon. Good cond. New tyres. 42 000 km.

8 Calculate the difference in price between the cars.

a
Colt	$8 905
BMW	$5 905

b
Barina	
Colt	

c
Barina	
Ford	

d
Barina	
Toyota	

e
Ford	
Toyota	

f
Colt	
Toyota	

9 Mr Samuels bought 2 of the cars above for his business. Which cars did he buy if the difference in their price was $285?

_____

N3.2 Students identify and solve addition and subtraction problems involving whole numbers, and decimal fractions in context, selecting from a range of computation methods, strategies and known number facts.

125

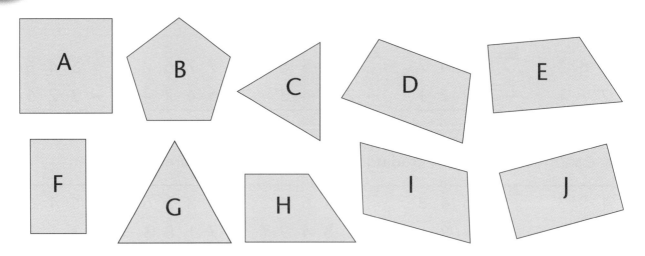

10   Group the 2D shapes by writing the correct letters after each question.

a   Shapes with right angles   _____

b   Shapes that do not have right angles   _____

c   Shapes that are equilateral triangles   _____

d   Shapes with angles smaller than a right angle   _____

e   Shapes that are quadrilaterals   _____

f   Shapes that are trapeziums   _____

g   Shapes that are parallelograms   _____

h   Shapes with right angles and parallel sides   _____

i   Shapes that will fold exactly in half   _____

11   Draw examples of each shape. Their bases must not be parallel to the bottom of the page. The first one has been done as an example for you.

a Square	b Rhombus	c Rectangle
d Right angle triangle	e Trapezium	f Scalene triangle

S3.1 Students describe the defining geometric properties of families of 3D shapes, model 3D shapes using nets and other representations, and identify and describe the properties of specific families and subgroups of 2D shapes.

Timetables

CITY TO BEENLEIGH
Early afternoon trains

Central	12:05	12:35	1:05
Roma Street	12:09	12:39	1:09
South Brisbane	12:13	12:43	1:13
South Bank	12:15	12:45	1:15
Park Road	12:17	12:47	1:17
Loganlea	12:54	1:24	1:54
Beenleigh	1:03	1:33	2:03

12 Interpret the timetable.

a What time does the 12:05 train from Central arrive at Beenleigh? _____

b What time does the 12:35 train from Central arrive at Loganlea? _____

c What time does the 1:05 train from Central arrive at South Bank? _____

d What time does the 12:35 train from Central arrive at South Brisbane? _____

13 How long does it take the trains to travel from:

a Central to Roma Street? _____ c Roma Street to South Bank? _____

b Central to South Brisbane? _____ d Loganlea to Beenleigh? _____

14 If you caught the 12:35 from Central where would you be at these times?

a 1:24 _____ b 12:43 _____ c 1:33 _____

15 How long is it between trains from Central? _____

16 Record the time when you do these things. Use digital time.

Event	Time
Get out of bed	☐☐ : ☐☐
Have breakfast	☐☐ : ☐☐
Arrive at school	☐☐ : ☐☐
Start first lesson	☐☐ : ☐☐
Start recess	☐☐ : ☐☐
Finish recess	☐☐ : ☐☐
Start lunch	☐☐ : ☐☐

M3.2 Students read, record and calculate with 12-hour time, and interpret calendars and simple timetables related to daily activities.

127

Cashless transactions

EFTPOS cards allow people to carry out cashless transactions. By swiping the card at an Automatic Teller Machine (**ATM**) money is transferred electronically from one account to another.

1 Study this card in order to answer the questions.

WITCH BANK ◆

1234 5678 9012 3456

Expires 09/15
Ms Sava Lotte   *OOO PLUS*

a Name the bank that issued the card.

_____

b Name the bank customer. _____

c Give the full date upon which the card expires. _____

d How many digits are in the identification number? _____

2 Examine Ms Sava Lotte's bank statement to answer the questions below.

WITCH BANK — Transaction statement

Account Number 1234 5678 9012 3456   **Period 1 July to 31 July**

Ms Sava Lotte

Date	Details	Withdrawals	Deposits	Balance
01-07-10	Opening balance			$9 999.00
04-07-10	Australia Post	$25.00		$9 974.00
09-07-10	Myer	$125.00		$9 849.00
15-07-10	Bundaberg Sports Club	$50.00		$9 799.00
18-07-10	Shell Service Station	$44.00		$9 755.00
18-07-10	Myer	$100.00		$9 655.00
18-07-10	Bundy Real Estate	$440.00		$9 215.00
24-07-10	Payment		$5 000	
28-07-10	Myer	$60.00		

a What period of time does the statement cover? _____

b How many transactions were made altogether? _____

c What was the total amount spent at Myer? _____

d How much was deposited on July 24? _____

e How much was spent on July 18? _____

f Complete the balance column for July 24 and July 28.

N3.1 Students compare, order and represent whole numbers to 9 999 and common and decimal fractions, calculate cash transactions and describe other methods of payment.

Hardly Normal Furniture and White Goods

$1 425

$3 165

$2 172

$2 599

$3 570

3 Use any strategy you wish to solve these problems.

a	How much would it cost Ben to buy a TV and a lounge?	
b	How much would it cost Erica to buy a refrigerator and a dishwasher?	
c	How much more does Ryan need to save if he wants the TV but only has $1 878?	
d	How much did Hannah pay for the refrigerator if she was given a discount of $299?	

e	If Tom was given $199 cash back for his old TV, how much did he pay for a new one?	
f	How much would it cost Fiona to buy the dishwasher and the washing machine?	
g	What is the difference in price between the washing machine and the dishwasher?	
h	How much would it cost to buy the lounge, TV and refrigerator?	

4 Complete the number cross.

1		2			3	4			5
	6		7			8	9		
10			11						
12					13				
			14						

Across

1 356 + 103 =
3 580 + 400 =
6 367 + 535 =
8 25 + 53 =
11 407 + 380 =
12 7 + 5 =
13 587 + 155 =
14 42 + 54 =

Down

2 500 + 400 =
4 687 + 200 =
5 276 + 118 =
7 144 + 134 =
9 559 + 275 =
10 347 + 268 =

N3.2 Students identify and solve addition and subtraction problems involving whole numbers, and decimal fractions in context, selecting from a range of computation methods, strategies and known number facts.

129

5 Jordan was watching the bricklayer laying pavers and noticed the patterns. Complete each pattern.

a

b

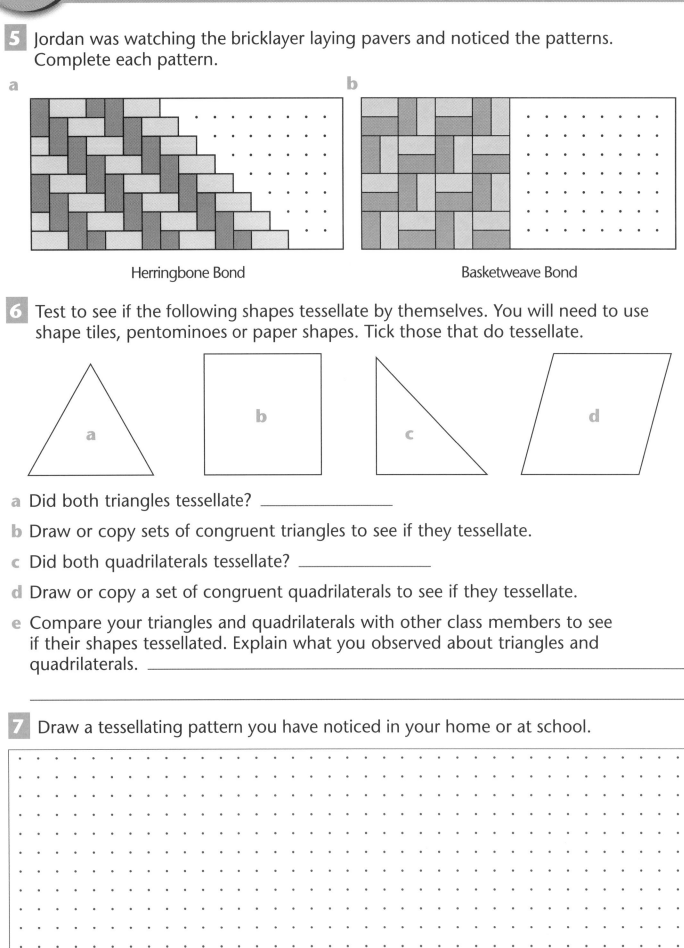

Herringbone Bond   Basketweave Bond

6 Test to see if the following shapes tessellate by themselves. You will need to use shape tiles, pentominoes or paper shapes. Tick those that do tessellate.

a   b   c   d

a Did both triangles tessellate? _____

b Draw or copy sets of congruent triangles to see if they tessellate.

c Did both quadrilaterals tessellate? _____

d Draw or copy a set of congruent quadrilaterals to see if they tessellate.

e Compare your triangles and quadrilaterals with other class members to see if their shapes tessellated. Explain what you observed about triangles and quadrilaterals. _____

_____

7 Draw a tessellating pattern you have noticed in your home or at school.

S3.1 Students describe the defining geometric properties of families of 3D shapes, model 3D shapes using nets and other representations, and identify and describe the properties of specific families and subgroups of 2D shapes.

Data investigation

The local council conducted a survey to find the size of a typical family.

Murray	Burns	Jones	North	Hooks	Otford	Quinn
5	2	6	3	4	6	4

Ryan	Salem	Stuart	Smith	Vidler	Walsh	Talko
5	3	4	4	4	6	3

8 Make a table of the number of families with 2, 3, 4, 5 or 6 family members.

Family size	2	3	4	5	6
Number of families	1				

9 Use the data you have gathered on your table to create a bar graph that shows the number of families of 2, 3, 4, 5 or 6.

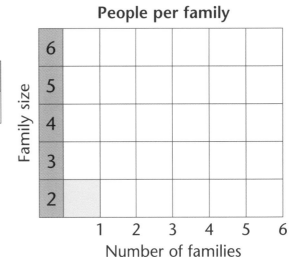

People per family

10 Conduct a survey of your own about the size of families in your school. Use the space below to record your data in a table.

11 Make statements about any patterns or trends you noticed in your data.

_____

_____

12 Who might make use of this kind of information? Why?

_____

_____

13 How could the data be useful for your school principal?

_____

_____

CD3.2 Students design and trial a variety of data collection methods and use existing sources of data to investigate their own and others' questions, organise data and create suitable displays identifying and interpreting elements of the displays.

131

3-digit × 1-digit multiplication

Problem
Tom had 4 stacks of cards with 124 cards in each stack.

Hund	Tens	Ones
	1	
1	2	4
×		4
4	9	6

4 ones × 4 = 16 ones. Place 6 in the ones column and trade the ten ones to make 1 ten.

1 hundred × 4 = 4 hundreds. Write 4 in the hundreds column.

2 tens × 4 = 8 tens plus the 1 traded equals 9 tens. Write 9 in the tens column.

1 Complete these multiplications mentally and check your answers with a calculator.

a

Hund	Tens	Ones
1	2	6
×		2

b

Hund	Tens	Ones
1	2	8
×		3

c

Hund	Tens	Ones
2	3	2
×		4

d

Hund	Tens	Ones
2	4	7
×		5

e

Hund	Tens	Ones
2	5	3
×		4

f

Hund	Tens	Ones
1	3	8
×		5

g

Hund	Tens	Ones
2	5	4
×		6

h

Hund	Tens	Ones
3	2	3
×		7

i

Hund	Tens	Ones
4	2	0
×		8

j

Hund	Tens	Ones
1	3	7
×		6

k

Hund	Tens	Ones
2	1	6
×		6

l

Hund	Tens	Ones
3	2	5
×		4

m

Hund	Tens	Ones
2	0	8
×		7

n

Hund	Tens	Ones
1	7	9
×		3

o

Hund	Tens	Ones
3	5	6
×		8

2 Calculate the following masses.

SPAGHETTI 130g

Jalapeno DIP 350g

CUPPA COFFEE 150g

WAWA Baked Beans 225g

a The mass of 4 tins of spaghetti.

b The mass of 6 containers of dip.

c The mass of 8 jars of coffee.

d The mass of 8 cans of beans.

N3.3 Students identify and solve multiplication and division problems involving whole numbers, and decimal fractions in context, selecting from a range of computation methods, strategies and known number facts.

Multiplying and dividing decimals

3 Multiply these decimals by 10. Use your calculator for **a**, **b**, **c** and **d** and do the others mentally.

a 0.45×10 [ ]

b 0.23×10 [ ]

c 0.46×10 [ ]

d 0.52×10 [ ]

e 3.16×10 [ ]

f 4.17×10 [ ]

g 5.28×10 [ ]

h 7.33×10 [ ]

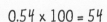
$0.54 \times 10 = 5.4$

4 What did you notice when you multiplied the decimals by 10?

_____

5 Multiply these decimals by 100. Use your calculator for **a**, **b**, **c** and **d** and do the others mentally.

a 0.25×100 [ ]

b 0.32×100 [ ]

c 0.64×100 [ ]

d 0.78×100 [ ]

e 6.23×100 [ ]

f 7.14×100 [ ]

g 8.25×100 [ ]

h 3.76×100 [ ]

$0.54 \times 100 = 54$

6 What did you notice when you multiplied the decimals by 100?

_____

7 Divide the first four decimals using a calculator and do the others mentally.

a $5.26 \div 10$ [ ]

b $8.49 \div 10$ [ ]

c $9.86 \div 10$ [ ]

d $7.21 \div 10$ [ ]

e $85.6 \div 10$ [ ]

f $99.9 \div 10$ [ ]

g $25.3 \div 10$ [ ]

h $72.8 \div 10$ [ ]

I can do $99.9 \div 10$ in my head.

8 What did you notice when you divided the decimals by 10?

_____

9 Divide the first four decimals using a calculator and do the others mentally.

a $85.6 \div 100$ [ ]

b $99.9 \div 100$ [ ]

c $25.3 \div 100$ [ ]

d $68.7 \div 100$ [ ]

e $32.5 \div 100$ [ ]

f $90.4 \div 100$ [ ]

g $180 \div 100$ [ ]

h $275 \div 100$ [ ]

i $368 \div 100$ [ ]

10 What did you notice when you divided the decimals by 100?

_____

N3.3 Students identify and solve multiplication and division problems involving whole numbers, and decimal fractions in context, selecting from a range of computation methods, strategies and known number facts.

133

11 Colour each object and its net the same colour. Make sure you use a different colour for each pair.

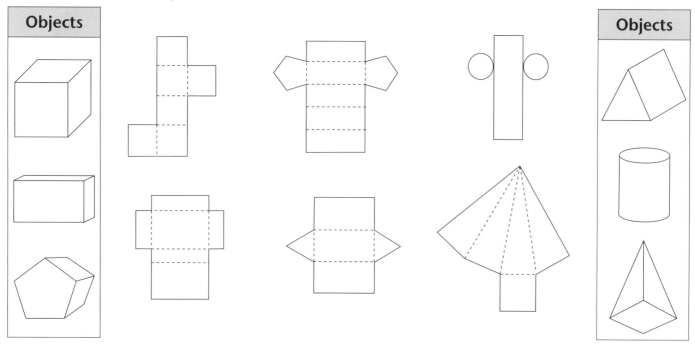

12 Use polydrons or paper shapes to discover which nets fold to make a closed 3D shape. Use the grid below to record your answers.

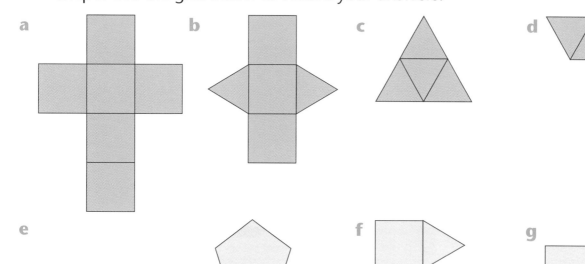

	Forms a 3D object					
Forms a 3D object						
Doesn't form a 3D object						

S3.1 Students describe the defining geometric properties of families of 3D shapes, model 3D shapes using nets and other representations, and identify and describe the properties of specific families and subgroups of 2D shapes.

Metres and centimetres

13 Use decimal notation to express the heights of these people in metres rather than metres and centimetres. The first one is done for you.

	Name	Height metres and centimetres	Decimal notation metres
a	Ryan	1 metre 37 centimetres	1.37 m
b	Bronte	1 metre 48 centimetres	
c	Dylan	1 metre 52 centimetres	
d	Cameron	1 metre 69 centimetres	
e	Sophie	1 metre 23 centimetres	
f	Mr Tallis	1 metre 95 centimetres	

1 metre 37 centimetres is the same as 1.37 metres.

14 Measure and record your own height. [ ]

15 Use a metre rule or a tape measure to record the lengths of the items in metres and centimetres as well as in decimal notation e.g. 1 m 63 cm = 1.63 m.

Larger distances can be measured with a trundle wheel.

	Distance	m and cm		Decimal
a	The length of a cupboard	m	cm	m
b	The length of a blackboard	m	cm	m
c	The width of your classroom	m	cm	m
d	The length of your classroom	m	cm	m
e	The length of a handball court	m	cm	m
f	The length of a car	m	cm	m

16 Place each set of measurements in ascending order.

a 1.55 m, 240 cm, 1 m 25 cm, 2.07 m _____

b 256 cm, 1.55 m, 1.05 m, 1 m 15 cm _____

c 1 m 45 cm, 208 cm, 1 m 10 cm, 2.10 m _____

d 2.65 m, 2.56 m, 2 m 15 cm, 2.04 m _____

e 5.50 m, 5 m 01cm, 4.99 m, 0.99 m _____

17 What is the difference in height between Mr Reilly who is 1.85 m and Sam who is 1 metre 58 cm tall?

Ascending order means going from lowest to highest.

_____

Division

1 Complete the division wheels.

a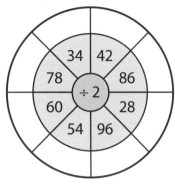

b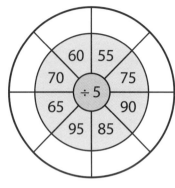

c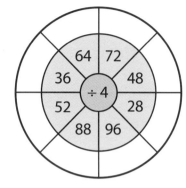

2 Work out the unknown numbers in these divisions.

a $3 \overline{) 5 \bigcirc}$ = 1 7

b $5 \overline{) 7 5}$ = $\bigcirc$ 5

c $\bigcirc \overline{) 9 1}$ = 1 3

d $4 \overline{) 6 \bigcirc}$ = 1 7

e $3 \overline{) 7 2}$ = $\bigcirc$ 4

f $\bigcirc \overline{) 8 4}$ = 1 2

g $2 \overline{) 3 \bigcirc}$ = 1 7

h $5 \overline{) 6 5}$ = $\bigcirc$ 3

i $\bigcirc \overline{) 9 6}$ = 2 4

j $3 \overline{) 7 \bigcirc}$ = 2 6

k $3 \overline{) 4 8}$ = $\bigcirc$ 6

l $6 \overline{) 7 \bigcirc}$ = 1 3

17 x 4 = 68 so 68 ÷ 4 = 17.

3 Complete the chart to show how much money was shared and the size of each share.

	Amount	Number of shares	Amount per share
a	$88		$44
b	$96	4	
c		8	$9
d	$90		$18
e	$80	10	
f	$99		$11
g		6	$16

If each share is $18 and there are 5 shares, I go $18 x 5 = $90.

4 Solve these problems using any strategy you wish.

a	How long is each section of the junior triathlon if the total distance of 42 km was broken into 6 equal sections?		
b	How much did Jacqui save each week if she saved the same amount each week and it took her 5 weeks to save $145?		

N3.3 Students identify and solve multiplication and division problems involving whole numbers, and decimal fractions in context, selecting from a range of computation methods, strategies and known number facts.

Adding and subtracting decimals

5 Add or subtract these decimal measurements.

a 3 4 . 2 6 m
  1 5 . 0 7 m
+ 1 6 . 2 3 m
_____

b 3 5 . 6 3 kg
−   7 . 5 8 kg
_____

c 1 4 . 7 9 m
   8 . 0 7 m
+ 1 5 . 6 3 m
_____

d 8 6 . 3 4 kg
− 3 4 . 5 6 kg
_____

e 3 0 . 8 6 m
  3 4 . 0 7 m
+ 2 4 . 6 8 m
_____

6 Solve the problems based on the map.

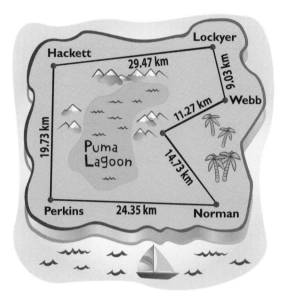

Hackett — 29.47 km — Lockyer

9.03 km

19.73 km

11.27 km — Webb

Puma Lagoon

14.73 km

Perkins — 24.35 km — Norman

a	What is the difference in the lengths of the trips from Hackett to Lockyer and from Perkins to Norman?	
b	Sam travelled from Norman to Lockyer via Jones and Webb. How many kilometres did he travel?	
c	Kelly travelled from Norman to Lockyer via Perkins and Hackett. How many kilometres did she travel?	
d	Sam said that his trip was shorter than Kelly's. How many kilometres shorter was Sam's trip?	
e	Tina left Lockyer and travelled a total distance of 59.38 km. What were the towns that Tina visited?	

7

Parcel A	Parcel B	Parcel C	Parcel D
88.17 kg	76.28 kg	67.77 kg	68.72 kg

Tina has a truck company and has four more parcels to deliver but only two trucks are left in the yard. Solve these problems for her.

a	This truck is almost full but can carry another 145 kg without breaking the legal limit. Which parcels would you put on this truck?	
b	This truck is also close to being full but can carry another 156 kg. Which parcels would you put on this truck?	
c	Tina charges $10 per kilogram to carry parcels. How much will she receive for delivering all four parcels?	

N3.2 Students identify and solve addition and subtraction problems involving whole numbers, and decimal fractions in context, selecting from a range of computation methods, strategies and known number facts.

137

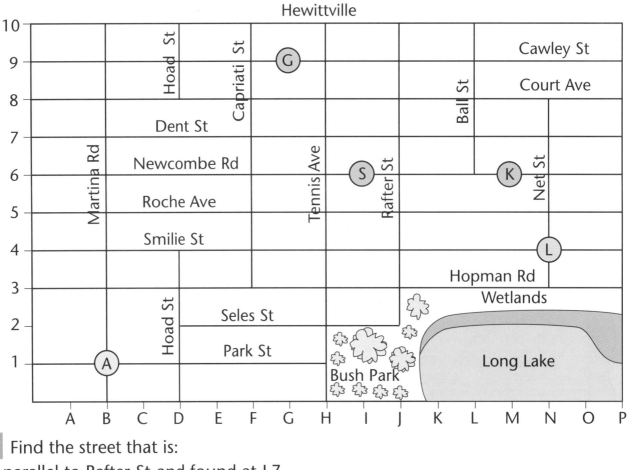

8 Find the street that is:

a parallel to Rafter St and found at L7 _____

b parallel to Dent St and found at G6 _____

c parallel to Seles St and found at F1 _____

d parallel to Park St and found at G8 _____

e parallel to Hopman Rd and found at A1 _____

9 What letter is found at:

a B1? _____   **b** M6? _____   **c** N4? _____   **d** G9? _____   **e** I6? _____

10 Colour the shortest route to get from K to G.

11 Write a set of directions describing how to get from K to G.

_____

_____

_____

12 Give as many sets of grid reference points as you can for Long Lake.

S3.2 Students interpret and create maps and plans using a range of conventions, describe locations and give directions using major compass points, angles and grids.

13 Experiment 1—to compare the volume of 3 objects

a Half-fill a clear measuring jug with water and mark the water level.

b Find 3 different shaped objects, such as stones, that can be submerged in your container and tie a piece of string to each one.

c Submerge the first object in the container and mark the water level before taking it out.

d Submerge the second object and mark the water level before taking it out.

e Submerge the third object and mark the water level before taking it out.

f Which object has the greatest volume? _____

g Discuss your experiment results with a friend and see if their results were similar to yours.

14 Experiment 2—to measure the amount of overflow

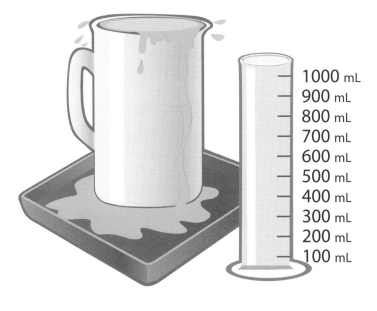

1000 mL
900 mL
800 mL
700 mL
600 mL
500 mL
400 mL
300 mL
200 mL
100 mL

a Place a container such as a jug on a tray.

b Fill the jug to the brim with water.

c Gently place a large rock or stone in the jug.

d Watch as water flows over the top of the jug and onto the tray.

e Measure the water in the tray by pouring it into a beaker. _____ mL

15 Which experiment do you think gave the more accurate measure of volume?

_____

_____

M3.1 Students identify and use equivalent forms of standard units when measuring, comparing and ordering, and estimate using a range of personal referents.

139

3-digit × 1-digit multiplication

1 Complete the multiplications. Circle the winning bingo card.

a

Hund	Tens	Ones
5	3	2
×		3

b

Hund	Tens	Ones
4	4	3
×		4

A

	948		1956
1596		3230	
	1772		3724

B

	3230		1956
1596		1836	
	1772		2226

Bingo!

c

Hund	Tens	Ones
6	4	6
×		5

d

Hund	Tens	Ones
2	3	7
×		4

e

Hund	Tens	Ones
3	2	6
×		6

f

Hund	Tens	Ones
5	3	2
×		7

g

Hund	Tens	Ones
2	4	2
×		8

h

Hund	Tens	Ones
4	3	7
×		6

i

Hund	Tens	Ones
1	4	8
×		8

2 Round each number to the nearest hundred to estimate the product before finding the exact answer.

a

Hund	Tens	Ones
2	9	8
×		2

est:

b

Hund	Tens	Ones
4	0	2
×		9

est:

c

Hund	Tens	Ones
4	9	2
×		8

est:

492 × 8
Think 500 × 8,
that's 4 000.

d

Hund	Tens	Ones
6	0	4
×		7

est:

e

Hund	Tens	Ones
6	8	9
×		5

est:

f

Hund	Tens	Ones
8	0	8
×		8

est:

N3.3 Students identify and solve multiplication and division problems involving whole numbers, and decimal fractions in context, selecting from a range of computation methods, strategies and known number facts.

Equivalence in equations

3 Supply a number sentence to balance those already on the balances. Your sentence must use the operation sign provided. The first one has been done for you.

a | 48 + 22 | 50 + 20

b | 98 − 82 | +

c | 20 × 5 | ×

d | 36 ÷ 6 | ÷

e | 85 − 50 | +

f | 17 + 15 | −

g | 15 × 2 | ÷

h | 90 ÷ 2 | ×

Order of operations
There are special rules that tell us the order in which to perform operations. They are:
- always do the work in the **brackets first** (4 + 5) × 2 = 18
- do operations with division and multiplication from **left to right**
  12 × 2 ÷ 6 = 4

4 Apply the order of operations rules to complete these equations.

a 3 × 4 ÷ 2 = ☐

b 20 − 5 + 3 = ☐

c 16 − 8 + 7 = ☐

d 21 ÷ 3 × 2 = ☐

e 4 × 5 ÷ 2 = ☐

f 20 ÷ 5 × 3 = ☐

g 12 ÷ 4 × 2 = ☐

h 4 × 9 ÷ 6 = ☐

i 16 ÷ 8 × 5 = ☐

j 25 × 4 ÷ 2 = ☐

4
20 ÷ 5 × 3 = 12

5 Use the ☐ = ☐ or ☐ ≠ ☐ signs to compare these number sentences.
Remember to follow the order of operations.

a (5 × 4) × 3 ☐ 5 × (4 × 3)

b 4 × (3 × 2) ☐ (4 × 3) × 2

c 95 + 5 − 20 ☐ 50 + 20 + 5

d 75 − 30 + 15 ☐ 90 − 30

e (24 ÷ 6) × 2 ☐ 160 ÷ 20

f 6 × (18 ÷ 2) ☐ (3 × 9) × 2

g 60 − 15 + 12 ☐ 23 + 14 − 5

h (7 × 8) ÷ 2 ☐ 2 × 4 × 7

i 18 − 9 + 18 ☐ (35 − 18) + 20

j (24 ÷ 4) × 2 ☐ (30 ÷ 5) × 3

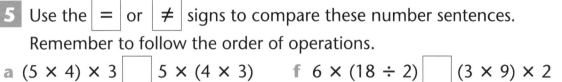

27
(3 × 9) × 2

PA3.2 Students represent and describe equivalence in equations that involve combinations of multiplication and division or addition and subtraction.

141

Remember!
A diagonal is a line that joins two non-adjacent vertices.
Congruent shapes are identical, that is
the same shape and size.
A line of symmetry creates a mirror image so
that both sides match each other exactly.

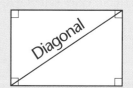

Diagonal

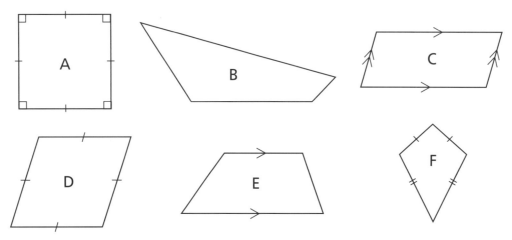

A   B   C

D   E   F

6 Give the letters of the quadrilaterals above that share these properties.

	Property	Shapes				
a	Opposite sides are parallel					
b	All sides are the same length					
c	Two sets of parallel sides					
d	Four right angles					
e	No sides the same length					

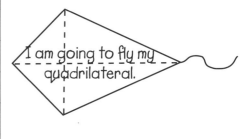

I am going to fly my quadrilateral.

7 Draw a diagonal on the quadrilaterals above and colour the shapes that produce congruent triangles. Use two colours.

8 Read the descriptions below and then draw and name each shape.

My diagonals are also lines of symmetry. All my sides are the same length. I don't have any right angles.	My opposite sides are equal and parallel. I have four right angles.

S3.1 Students describe the defining geometric properties of families of 3D shapes, model 3D shapes using nets and other representations, and identify and describe the properties of specific families and subgroups of 2D shapes.

Gympie Cricket Club spreadsheet

	A	B	C	D
1	Date	Item	Cost	Balance
2	May 1	Opening balance		$900
3	May 4	Pads	$120	$780
4	May 7	Helmets	$240	$540
5	May 19	Bats	$300	$240
6	May 23	Gloves	$ 60	$180
7	May 25	Balls	$ 90	$ 90
8				$

Spreadsheets are used to organise and display data.

9 How much money did Gympie Cricket Club spend on these items?

a Pads _____   **b** Balls _____   **c** Bats _____

d Helmets _____   **e** Gloves _____

10 Discuss with a friend how the balance column works then write an explanation.

_____

_____

11 On 30 May, Gympie Cricket Club spent $50 on insurance. Add this to item 8 on the spreadsheet above.

12 Complete the balance column on the Baker family budget spreadsheet.

Subtract the cost of the item from the previous balance to find the new balance.

The Baker family budget

	A	B	C	D
1	Date	Item	Cost	Balance
2	Aug 3	Opening balance		$600
3	Aug 4	Groceries	$150	$450
4	Aug 5	Fruit & veg	$ 50	$400
5	Aug 8	Bills	$100	$
6	Aug 12	Meat	$ 80	$
7	Aug 16	Car	$ 60	$
8	Aug 20	Entertainment	$ 80	$

N3.2   CD3.2 Students design and trial a variety of data collection methods and use existing sources of data to investigate their own and others' questions, organise data and create suitable displays identifying and interpreting elements of the displays.

143

Contracted multiplication

1 Complete the multiplications. The first one is done for you.

a

Hund	Tens	Ones
1	1	
1	2	3
×		5
6	1	5

b

Hund	Tens	Ones
2	2	8
×		4

c

Hund	Tens	Ones
2	3	4
×		3

d

Hund	Tens	Ones
1	2	6
×		6

e

Hund	Tens	Ones
1	1	9
×		5

f

Hund	Tens	Ones
1	2	5
×		5

g

Hund	Tens	Ones
1	3	4
×		6

h

Hund	Tens	Ones
1	2	8
×		7

i

Hund	Tens	Ones
1	3	6
×		8

j

Hund	Tens	Ones
1	4	3
×		9

2 Find the missing numbers in the multiplications below.

a

Hund	Tens	Ones
3	3	
1	6	7
×		☐
8	3	5

b

Hund	Tens	Ones
3	2	
1	7	6
×	☐	
7	0	4

c

Hund	Tens	Ones
2	4	
1	3	☐
×		6
8	2	8

d

Hund	Tens	Ones
	3	
1	☐	5
×		6
6	9	0

e

Thou	Hund	Tens	Ones
	7	4	
	1	9	☐
×			8
1	5	6	8

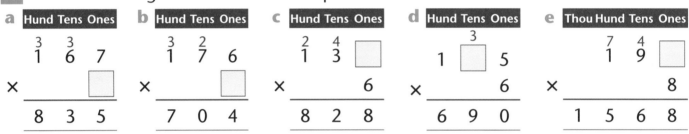

$124.95 $132.99 $104.88 $129.90 $109.50

3 Round each price to the nearest dollar in order to work out the approximate cost of these items.

a	Approximately how much would it cost Mrs Johnson to buy 3 MP3 players, one each for her son, niece and nephew?	
b	If Sunshine Real Estate is buying 4 new printers, about how much will it cost them?	
c	The music department at Merton High School is planning on buying 5 DVD players. Approximately how much will it cost them?	
d	About how much will it cost the owners of Mack's Motel to buy 8 new televisions?	
e	The *Daily News* is buying 6 digital cameras for their reporters. About how much will they have to spend?	

N3.3 Students identify and solve multiplication and division problems involving whole numbers, and decimal fractions in context, selecting from a range of computation methods, strategies and known number facts.

Decimal number patterns

4 Complete the decimal counting patterns.

a
0.1	0.2	0.3			

d
0.23	0.25	0.27			

b
0.2	0.4	0.6			1.2

e
1.25	1.30	1.35			

c
0.3	0.6	0.9	1.2		

f
1.67	1.70	1.73			

5 Use the constant addition function on your calculator to make the following patterns.

a $1 + 0.5 = 1.5 = 2 = 2.5 = \square = \square = \square$

b $1.5 + 0.1 = \square = \square = \square = \square = \square = \square$

c $2.3 + 0.2 = \square = \square = \square = \square = \square = \square$

d $1.8 + 0.3 = \square = \square = \square = \square = \square = \square$

e $2.3 + 0.6 = \square = \square = \square = \square = \square = \square$

f $2.6 + 0.8 = \square = \square = \square = \square = \square = \square$

g $2.6 + 2.2 = \square = \square = \square = \square = \square = \square$

6 Follow the rules to find the output of the machines.

a
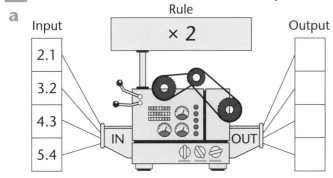

Input: 2.1, 3.2, 4.3, 5.4 — Rule × 2 — Output

c
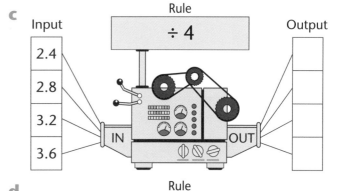

Input: 2.4, 2.8, 3.2, 3.6 — Rule ÷ 4 — Output

b
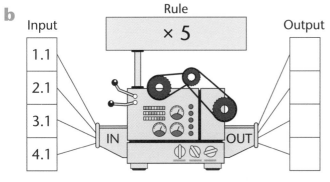

Input: 1.1, 2.1, 3.1, 4.1 — Rule × 5 — Output

d
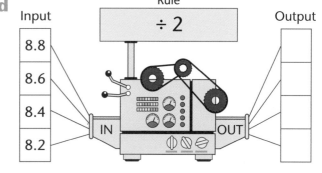

Input: 8.8, 8.6, 8.4, 8.2 — Rule ÷ 2 — Output

7 Draw these 3 shapes on a computer, then combine them into a design.

8 Choose any 3 shapes from below, then combine them on your computer to form a design. Sketch your design in the box.

9 Draw a triangle like the one shown here, then flip it and turn it 90° clockwise on the computer screen. Draw your results.

a flip

b turn

10 Draw a congruent copy of each shape on the dot paper.

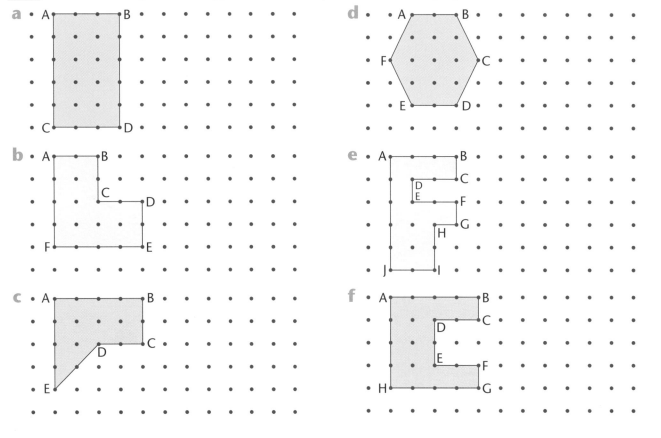

S3.1 Students describe the defining geometric properties of families of 3D shapes, model 3D shapes using nets and other representations, and identify and describe the properties of specific families and subgroups of 2D shapes.

Timelines show a span of time. They are broken up into equal intervals so events can be recorded on them.

Will's morning

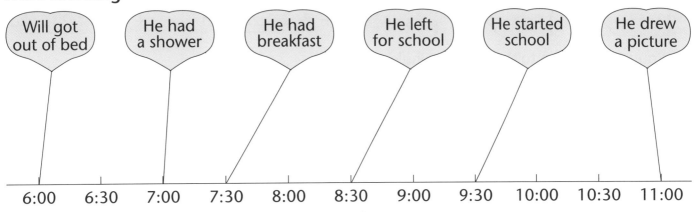

| Will got out of bed | He had a shower | He had breakfast | He left for school | He started school | He drew a picture |

6:00   6:30   7:00   7:30   8:00   8:30   9:00   9:30   10:00   10:30   11:00

11 What time did Will do these things?

a Have a shower   _____   d Have breakfast   _____

b Leave for school   _____   e Start school   _____

c Draw a picture   _____   f Get out of bed   _____

12 Draw a line to match the events in Sally's life to a place on the timeline.

| Sally was born in 1998 | She started walking in 1999 | She started pre-school in 2001 | She started school in 2003 | She started netball in 2006 | She turned 10 in 2008 |

1998   1999   2000   2001   2002   2003   2004   2005   2006   2007   2008   2009

13 Draw a line to match the times Tom did these things to a place on the timeline.

| Was a fool on 1 April | Had birthday party on 7 April | Went to Easter Show 10 April | Played soccer 17 April | Slept in 24 April | Athletics carnival 26 April |

1 2 3 4 5 6 7 8 9 10 11 12 13 14 15 16 17 18 19 20 21 22 23 24 25 26 27 28 29 30

Days in April

M3.2 Students read, record and calculate with 12-hour time, and interpret calendars and simple timetables related to daily activities.

147

Diagnostic review 4

PART 1

$19.99 $24.50 $15.25

a Round each price to the nearest whole dollar to work out an approximate price for all 3 books.

_____

b Calculate the total cost of the books and round it to the nearest 5 cents.

$ _____

$ _____

$ _____

$ _____

$ _____

c How much change would I receive from $60? _____

PART 2
N3.3

Complete the algorithms and problems.

a   2 6   **b**   3 5   **c**   5 6
×   5   ×   6   ×   8
_____   _____   _____

d 1 4 8   **e** 3 4 7   **f** 2 1 9
×   4   ×   5   ×   3
_____   _____   _____

g
5⟌25

h
6⟌72

i
7⟌91

j
3⟌315

k
5⟌750

l
8⟌968

m The 7 grades at Woy Wong Primary School have 130 students in each. How many students are there in the primary school? _____

PART 3
N3.1

Add or subtract the decimals.

a   1 6 . 3 4 m
1 3 . 3 8 m
+ 2 2 . 2 4 m
_____

b   4 3 . 0 4 kg
2 4 . 6 3 kg
+ 1 2 . 2 5 kg
_____

c   9 4 . 8 5 kg
− 3 2 . 6 3 kg
_____

d   8 6 . 3 7 m
− 2 7 . 4 2 m
_____

PART 4
N3.2

Round these numbers to the nearest hundred in order to estimate the answer.

a $619 + 380 \approx$ ___   **b** $235 + 172 \approx$ ___

c $865 - 479 \approx$ ___   **d** $908 - 595 \approx$ ___

e
Robert had a balance of $5 872 in his bank account until he spent $2 908. Approximately how much does he have left? [____]

PART 5
PA3.1

Decide if these sentences are true or false.

a $5 \times 4 + 8 = 4 \times 7$ _____

b $18 \div 3 + 20 > 4 \times 5 \times 8$ _____

c $1\frac{1}{2} \times 12 < \frac{1}{2} \times 24 + 3$ _____

d $8 \times 6 \div 4 = 3 + 2 \times 6$ _____

PART 6
PA3.1

Continue the counting patterns.

a

1.2	2.4	3.6			

b

9.5	9.2	8.9			

Diagnostic review 4

PART 7
S3.2

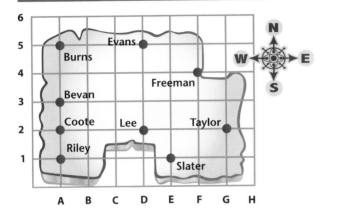

What towns are found at these grid references?

a A1 _____   b G2 _____

Give a set of references for:

c Burns _____   d Evans _____

e Bevan _____   f Lee _____

Give the direction from Lee to each town.

g Coote _____   h Evans _____

i Taylor _____

PART 8
S3.1

Draw a line to match the nets to the objects.

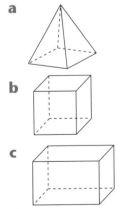
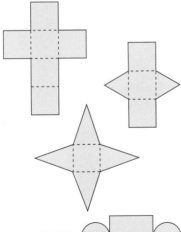
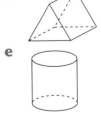
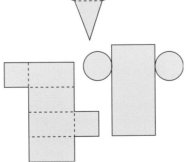

PART 9
M3.1

Calculate the distance around the outside boundary of these shapes.

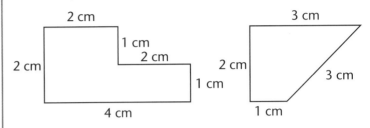

Distance = _____ cm   Distance = _____ cm

PART 10

How many square metres would be needed to cover a mat that is a square with 2 m sides?

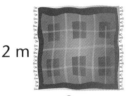
2 m

2 m

_____ m²

PART 11
M3.2

Complete the last column of the train timetable.

City to Beenleigh trains			
Central	1:35	2:05	2:35
Roma St	1:39	2:09	
South Bank	1:45	2:15	
Park Road	1:47	2:17	
Beenleigh	2:33	3:03	

PART 12
M3.1

Sam filled two containers to the 400 mL mark then placed a rock in each of them.

How much more water was displaced by the second rock than the first?

_____ mL

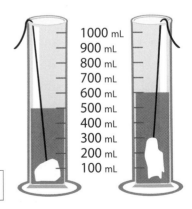

Juggling numbers

Cleo the clown is juggling five balls.
Each ball has a number.

Task 1:

Arrange the balls so that the total of each line (across and down) matches the number in the box below.

a

b

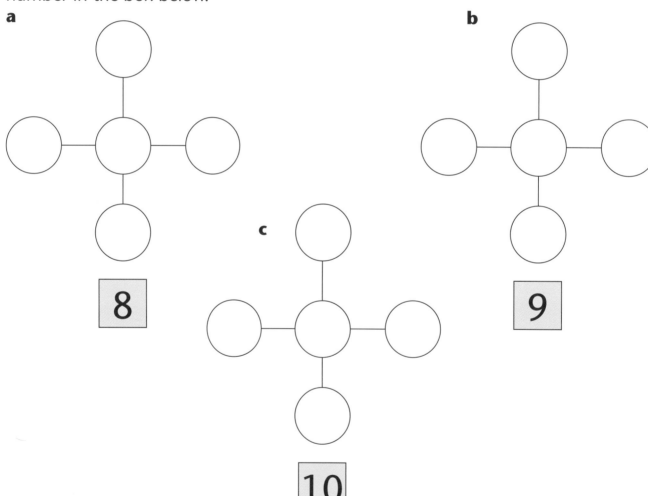

c

8

9

10

Task 2:

On another piece of paper, show how many different ways you can arrange the balls so that they make a total of 8.

Pizza supreme

Polly's Pizza Parlour sells four types of pizza by the slice—mushroom, salami, chicken and tropical. Ava, Riley, Oscar, Jade, Luke and Madison can choose four pizza slices each.

Mushroom	Salami	Chicken	Tropical
		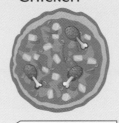	
$3 per slice	$2.40 per slice	$3.60 per slice	$3.30 per slice

Task:

Name the four pizza slices that each child may have bought.

 Ava spent $11.70.

_____ _____ _____ _____

 Riley also spent $11.70 but his order was different from Ava's.

_____ _____ _____ _____

 Oscar spent $13.50.

_____ _____ _____ _____

 Jade also spent $13.50 but her order was different from Oscar's.

_____ _____ _____ _____

 Luke spent $12.30.

_____ _____ _____ _____

 Madison also spent $12.30 but her order was different from Luke's.

_____ _____ _____ _____

Wild Kidz Clothing Company

Hayley, Liz and Amy went to the Wild Kidz Clothing Company to buy some new clothes. Each girl had a different amount of money to spend.

$19.99 $24.99 $29.99 $34.99 $13.99 $29.99

Task:

Write or draw all the different combinations of tops and bottoms that each girl could buy with her money.

 Hayley = $50

 Liz = $55

 Amy = $49

A new home

Jacqui and Paul want to buy a house. Jacqui's annual wage is $50 000 and Paul's annual wage is $44 992. So far they have saved $21 450 between them. They usually save $880 per month.

Task 1:

Circle the house you think Jacqui and Paul can best afford.

Colonial: $187 500	Hollywood: $365 000	Sea Breeze: $206 500

3 bedrooms big backyard new kitchen $18 700 deposit $195 weekly repayments	5 bedrooms huge pool ocean views $36 500 deposit $450 weekly repayments	3 bedrooms near a school nice backyard $20 650 deposit $220 weekly repayments

Task 2:

Explain why Jacqui and Paul should buy that house, and not one of the others.

_____

_____

_____

_____

_____

Slam dunk

The Stingers basketball team scored 95 points in its game against the Sharks.

Task 1:

Use the following clues to work out how many points each player scored.

Tyler scored 9 points more than Logan. _____ points

Noah scored $\frac{1}{3}$ of the points that Zoe scored. _____ points

Zoe scored 5 more points than Logan. _____ points

Logan scored 19 points. _____ points

Mia scored twice as many points as Noah. _____ points

Basketball goals may be worth 1, 2 or 3 points:
1-point goal = penalty shot from the free-throw line
2-point goal = shot from inside the arc
3-point goal = shot from outside the arc

Task 2:

Now work out how each player could have scored their points.
You can use tally marks to help you.

Player	3 points	2 points	1 point	Total
Tyler				
Noah				
Zoe				
Logan				
Mia				

Six 3-point goals equals 18 points.

Always the same?

Lucia has discovered a number puzzle that always produces the same answer—no matter what the starting number is.

Cool!

Task 1:

Investigate Lucia's puzzle to see if the answer is always the same.

Steps	Numbers			
Think of a number	between 1 and 10	between 50 and 100	between 100 and 500	between 500 and 1 000
Add 6				
Double it				
Subtract 4				
Halve it				
Subtract your starting number				

Task 2:

Can you add any more steps to the puzzle and still keep the same answer?

_____

_____

_____

_____

Most common coin

Ryan counted the coins in his money box and noticed that he had more 20-cent coins than any other type of coin. At school he said, "The 20 cent coin is the most common coin."

Step 1:

Collect as many different coins as you can and see which is the most common coin. (You can work with a partner to help you.)

Step 2:

Use tally marks to record your results.

Coin	5	10	20	50	1 DOLLAR	2 DOLLARS
Tally						

Step 3:

Use your data to create a bar graph.

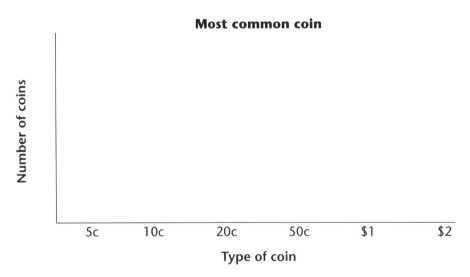

Step 4:

What change in circumstances could result in your data being different?

_____

_____

Food for thought

Cooper's family went to the Yummy Tummy Restaurant for dinner. They ordered three completely different meals for the same price.

Yummy Tummy Menu

Entrée	Main Meal	Dessert
Herb bread $5.50	Steak $18.50	Ice-cream $5.50
Soup $6.50	Chicken $19.50	Mud cake $7.00
Fish cakes $7.00	Fish $20.00	Pavlova $6.50
Salad $6.50	Pizza $18.50	Banana split $6.50
Pasta $6.00	Kangaroo $21.00	Pecan pie $7.00

What did Cooper's family order?

Task 1:

Select three meals, each with a different entrée, main meal and dessert. The total of each meal must add up to $31.50.

	Logan	Cooper	Thalia
Entrée			
Main			
Dessert			
Total	$31.50	$31.50	$31.50

Task 2:

Create your own meal for $30.00—it must include pasta.

_____

_____

_____

Court time

The Black and White Gymnasium Equipment Company has the job of marking lines on the school gym floor to create some multi-purpose courts. The standard size court has a length of 8 metres and a width of 6 metres.

The company must mark up a standard size court (A), a court with its dimensions halved (B), and a court with its dimensions doubled (C).

Mr Black said, "By doubling the dimensions we will double the area and by halving the dimensions we will halve the area."

Mrs White said, "Let's try it and see."

Task 1:

Draw the three courts on the grid below. Note: each square on the grid represents an area of 1 square metre.

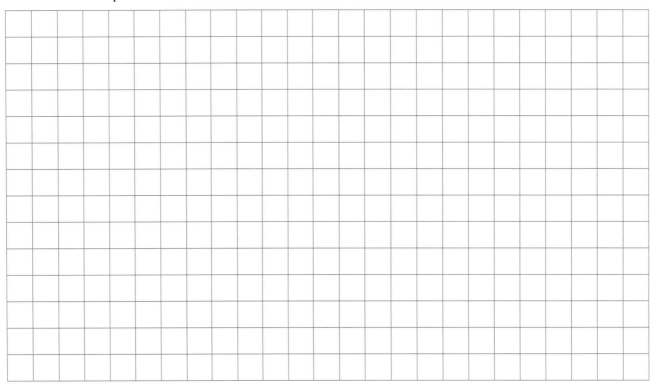

Task 2:

Calculate the area of each court in square metres.

A _____   B _____   C _____

Task 3:

What happened when the dimensions were doubled and halved?

_____

_____

Snakes and ladders

Investigation 10

The board game below is missing its snakes and ladders.

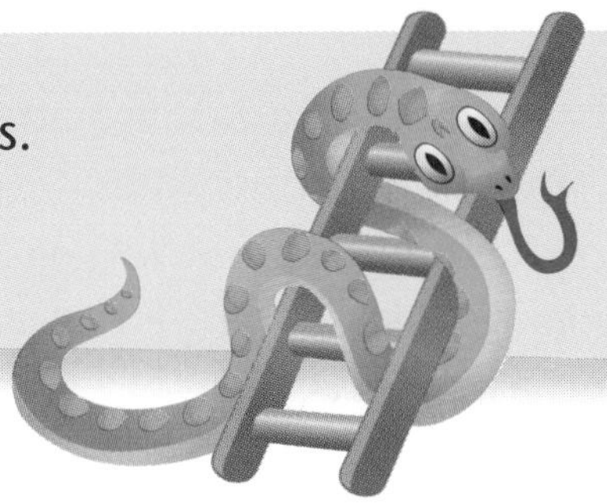

Task 1:

Add at least 5 snakes and 3 ladders to the game board. Make sure there is an equal chance of going forwards as backwards.

Snakes and Ladders—special edition created by ____________________

64	63	62	61	60 *Go back 3 spots*	59	58	57
49	50	51	52	53	54	55	56
48	47	46	45	44	43 *Have an extra throw*	42	41
33	34	35	36	37	38	39	40
32 *Jump ahead 5 spots*	31	30	29	28	27	26	25
17	18	19	20	21	22	23	24
16	15	14	13	12	11	10 *Miss a turn*	9
1	2	3	4	5	6	7	8

Task 2:

Explain where you placed your snakes and ladders and why.

Eyewitness

Three people had their bags stolen by a woman at a bus station. They could only remember that the woman was wearing glasses, large hoop earrings and a white T-shirt.

Help the police work out what the woman might have looked like.

Task:

Colour the mug shots below to show as many different appearances the woman could have, using the following features:

- Eye colour—blue, green or brown
- Hair colour—black, brown, red, blonde

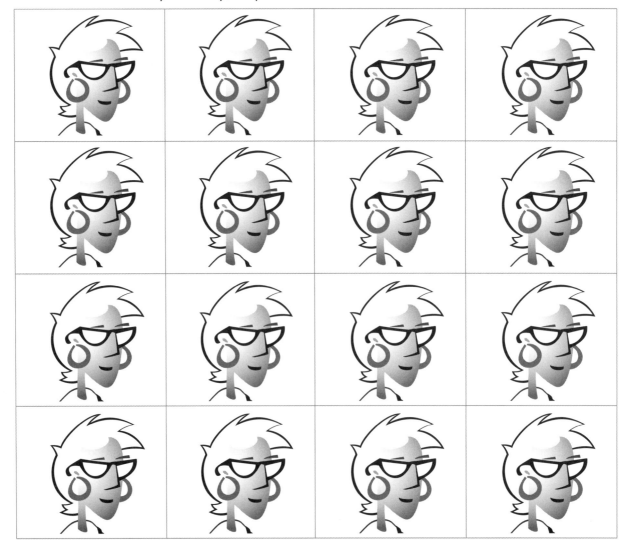

Time Flies

The table below shows the distances and flight times between Australian capital cities. All distances are shown but some flight times are missing.

Route	Kilometres	Hours: Minutes
Canberra–Sydney	240	0:45
Canberra–Melbourne	470	0:55
Hobart–Melbourne	620	
Adelaide–Melbourne	640	1:10
Sydney–Melbourne	710	1:10
Sydney–Brisbane	750	1:20
Canberra–Brisbane	950	
Canberra–Adelaide	970	1:30
Sydney–Adelaide	1170	1:40
Melbourne–Brisbane	1380	1:55
Adelaide–Brisbane	1620	2:15
Adelaide–Perth	2120	3:15
Adelaide–Darwin	2650	3:40
Perth–Darwin	2650	4:05
Perth–Melbourne	2710	
Brisbane–Darwin	2850	4:00
Melbourne–Darwin	3180	4:45
Perth–Sydney	3280	4:05

Task 1:

Use the data provided to estimate the missing times. Note: weather conditions, the size of the aeroplane and the size of its load are all factors that make it impossible to be entirely accurate with your estimates.

Task 2:

Plot and describe a course on the map below that involves 7 to 8 hours of flying time and covers a distance of about 5 500 kilometres.

_____

_____

_____

_____

_____

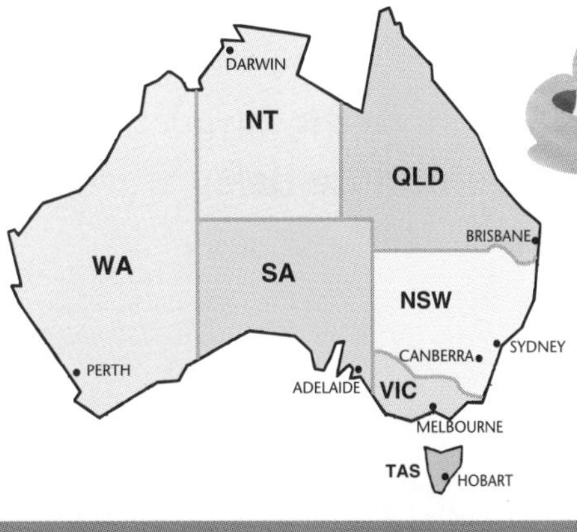

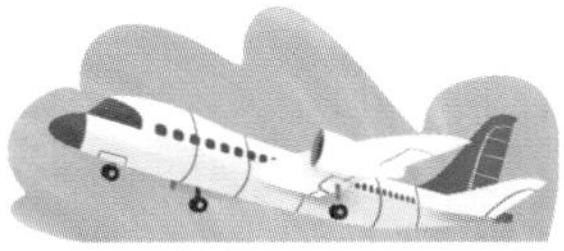

The next train ...

This is a copy of the train timetable for the Wilson line. Someone has spilt coffee on the timetable, blocking out some of the data.

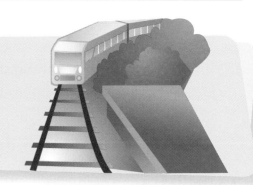

Train Timetable									
	1	2	3	4	5	6	7	8	9
Station	p.m.	p.m.	p.m.	p.m.	p.m.	p.m.	p.m.	p.m.	p.m.
Anderson	5:00	5:15	5:30	5:45	6:00	6:15	6:30	6:45	7:00
Brown		5:20		5:50		6:20		6:50	
Harris	5:10	5:25	5:40	5:55	6:10	6:25	6:40		
Jones		5:30		6:00		6:30			
King	5:20	5:35	5:50	6:05	6:20				
Lee		5:40		6:10					
Martin	5:30	5:45	6:00	6:15	6:30				
Ryan		5:50		6:20					
Smith	5:40	5:55	6:10	6:25	6:40	6:55		7:25	7:40
Stuart		6:00		6:30		7:00		7:30	
Wilson	5:55	6:05	6:20	6:35	6:50	7:05	7:20	7:35	7:50

Task 1:

Identify all the patterns you can see in the timetable.

_____

_____

_____

Task 2:

Add the missing times to the timetable.

Task 3:

In the grid below, show the times the trains would arrive at Stuart and Wilson if there was a twelve-minute delay starting at 5:40 p.m.

Stuart						
Wilson						

Paving patterns

Pamela's Paving Company supplies and lays paving bricks for backyards. Pamela likes to make interesting patterns with her paving bricks:

Popular paving patterns

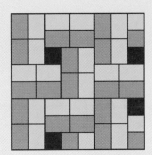  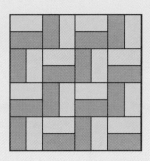

Task 1:

Use all three paving patterns to create interesting patterns in the yard.

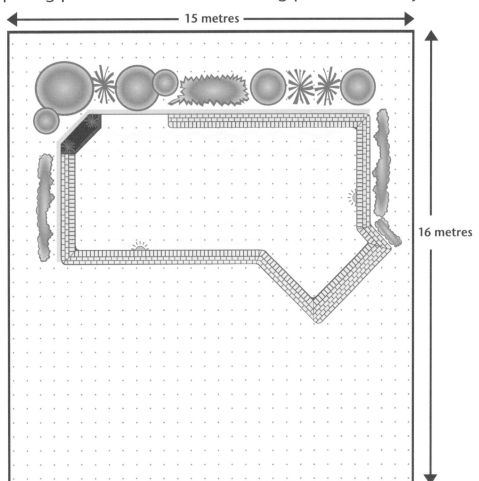

Task 2:

Estimate how many square metres of paving is required for this job.

Prize winner

Congratulations! You have won $5 999 worth of goods from the Bad Guys Department Store. You can select anything you like from the list below, but the total value must not be more than $5 999 or you will lose the lot.

$2 188 $598 $2 199 $129 $549

$349 $848 $1 499 $538

Task:

Create as many different combinations of goods you could select without exceeding $5 999.

_____

_____

_____

_____

_____

_____

_____

_____

_____

_____

The quickest way

Matilda is walking to school. Oliver is walking to softball practice.

Help Matilda and Oliver find the quickest way.

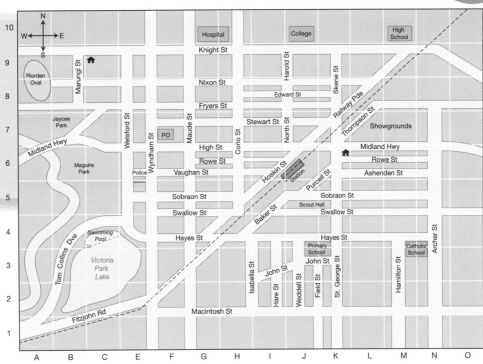

Task 1:

Describe the route Oliver should take to get from his home at K6 in Rowe Street to Riordan Oval.

_____

_____

_____

Task 2:

Describe the route Matilda should take to get from her home at C9 in Knight Street to the Catholic school.

_____

_____

_____

Task 3:

Using grid reference points, describe how a police car could travel from the showgrounds at L7 to the station in Hoskin Street and then to the police station.

_____

_____

_____

My investigation review

Name: _____ **Date:** _____

Investigation: _____

Before

1 What do I have to do? _____

_____

_____

2 Do I have a strategy? _____

_____

_____

3 Do I need materials? _____

_____

_____

4 How am I going to do it? _____

_____

_____

After

1 What did I find out? _____

_____

_____

_____

_____

_____

2 Did my strategy work? _____

_____

_____

_____

_____

_____

_____

© Oxford University Press 2009
This sheet may be copied for non-commercial classroom use.

Dictionary

abacus
An instrument used for calculating.

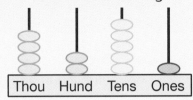

acute angle
An angle less than 90°.

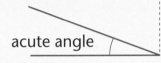

addition (+)
The operation that finds the sum or total.

a.m. (*ante meridiem*)
The morning. Any time from midnight to noon, e.g. 7:30 a.m. is 7:30 in the morning.

analogue clock
A clock face with numbers 1 to 12, and two hands.

angle
The amount of turn between two lines around a common endpoint (the vertex).

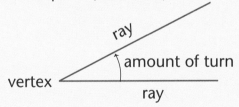

area
The surface covered by any 2D shape. Area can be measured in cm², m², hectares and km².

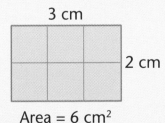

array
An arrangement of shapes that shows numbers.

ascending order
An arrangement of numbers from smallest to largest.

256, 291, 307, 452

associative property
A series of numbers can be added in any order without changing the result.

$$5 + 4 + 6 = 15$$
$$4 + 6 + 5 = 15$$
$$6 + 5 + 4 = 15$$

A series of numbers can be multiplied in any order without changing the result.

$$5 \times 4 \times 3 = 60$$
$$4 \times 3 \times 5 = 60$$
$$3 \times 5 \times 4 = 60$$

axis of symmetry
An imaginary line that divides a shape exactly in half. If a shape is folded along this line, both sides will match.

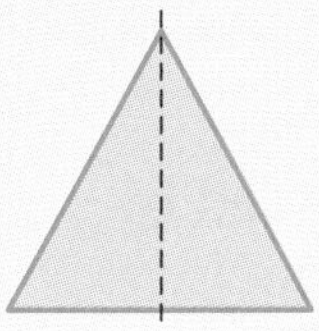

bar graph
A graph which uses vertical columns to represent data.

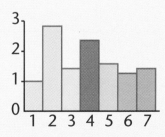

Dictionary

base
The bottom line of a 2D shape.

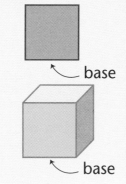

base

The bottom face of a 3D shape.

base

For example:
• pyramids have one base

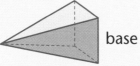

base

• prisms have two bases.

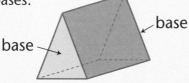

base

base

capacity
The amount a container can hold. Capacity can be measured in millilitres (mL), litres (L) and kilolitres (kL).

centimetre (cm)
A unit for measuring length. 100 cm = 1 metre.

circle
A plane shape bounded by a continual curve that is always the same distance from the centre point.

commutative property
Two numbers can be added in any order to give the same total.

$$15 + 13 = 28$$
$$13 + 15 = 28$$

Two numbers can be multiplied in any order to give the same product.

$$5 \times 4 = 20$$
$$4 \times 5 = 20$$

compass points
The cardinal compass points are north, south, east and west.

cone
A 3D shape with a circular base, tapering to a point (the apex).

corner (vertex)
The point where two or more lines meet to form an angle.

corner

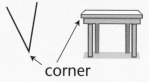

cross-section
The face that is left when a solid (3D) shape is cut through, parallel with its base.

cube
A 3D shape with six square faces, eight vertices and twelve edges.

cubic centimetre
A unit of volume. A centimetre cube has a volume equal to one cubic centimetre.

1 cm   1 cm

1 cm

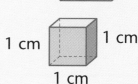

cylinder
A shape with two circular faces and one curved surface.

Dictionary

data
Information gathered together, such as a set of numbers or facts.

decade
Ten years.

e.g.   2001–2011

decimal
A fraction can be written as a decimal, e.g. 75 out of 100 can be written as 0.75 in decimal form.

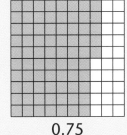

0.75

decimal point
A point used to separate the fraction part from the whole number.

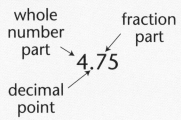

denominator
The bottom number of a fraction that tells how many parts there are in the whole.

$$\frac{1}{4} \begin{array}{l} \leftarrow \text{numerator} \\ \leftarrow \text{denominator} \end{array}$$

descending order
An arrangement of numbers from largest to smallest.

99, 76, 54

diagonal
A line which joins two non-adjacent corners of a polygon.

digital clock
A clock which displays only numbers. It has no hands.

dimension
A measurement of length, width or height.

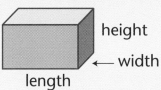

division (÷)
The operation that breaks groups or numbers into equal parts.

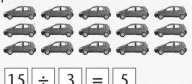

$$15 \div 3 = 5$$

double
Multiply by two.

edge
The intersection of two faces on a 3D shape.

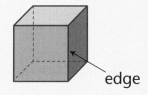

equivalent fractions
Fractions having the same value.

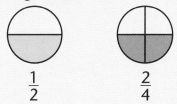

$$\frac{1}{2} \qquad \frac{2}{4}$$

faces
The surfaces of a 3D shape.

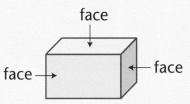

flip (reflect)
To turn a shape over.

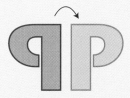

fraction
Any part of a whole.

$$= \frac{1}{3}$$

Dictionary

front view
The view we see when we look at an object from the front.

3D shape
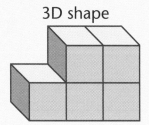

front view
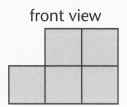

gram
A unit for measuring mass.

1000 grams = 1 kilogram

greater than (>)
The 'greater than' symbol shows the relationship between two unequal numbers.

8 > 5

grid reference points
Grid reference points show position on a grid. Grid reference points are read horizontally before vertically.

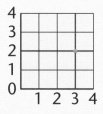

grouping
A way of dividing an amount into equal parts.

3 equal groups of 4 = 12

hexagon
A 2D shape with six sides.

regular hexagon

irregular hexagon

horizontal
At right angles to the vertical.

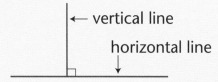
vertical line
horizontal line

hundredth
One part of a whole that has 100 parts altogether.

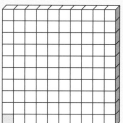

kilogram (kg)
The base unit for measuring mass.

1 kg = 1000 grams

kilometre (km)
A unit of length.

1 km = 1000 metres

length
The bigger of the two dimensions of a shape.

length
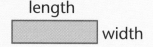
width

less than (<)
The 'less than' symbol shows the relationship between two unequal numbers.

5 < 8

line of symmetry
A line which divides a shape in half exactly. Shapes can have more than one line of symmetry.

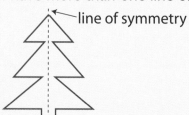
line of symmetry

litre (L)
A unit of capacity.

1 L = 1000 millilitres

mass
The amount of substance in an object.

1000 grams = 1 kg
1000 kg = 1 tonne

metre (m)
A unit of length.

1 metre = 100 cm

millilitre (mL)
A unit of capacity.
An object with a volume of 1 cm³ displaces 1 mL of water.

1000 mL = 1 litre

1 cm   1 cm
1 cm

millimetre (mm)
A unit of length.

10 mm = 1 centimetre

multiple
The result of multiplying a given number by any other number is a multiple of that given number.

Multiples of 4 are: 4, 8, 12, 16, 20, etc.

Multiples of 5 are: 5, 10, 15, 20, 25, etc.

multiplication (×)
The operation which finds the product of two or more numbers. Multiplication can be seen as repeated addition.

2 + 2 + 2 + 2 + 2 = 10
2 × 5 = 10

net
A flat shape that can be folded to make a 3D shape.

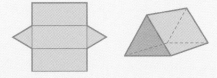

number line
A line on which numbers are marked. Number lines can be used to represent operations.

3 + 5 = 8

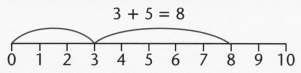

number pattern
Any set of numbers that follow a pattern or sequence.

1, 3, 5, 7, ___ , ___ , (The pattern is + 2)

3, 9, 27, ___ , ___ , (The pattern is × 3)

numeral
Any figure used to represent a number.

e.g. 0, 1, 2, 3, 4, 5

numerator
The top number of a fraction, telling us how many parts there are out of the whole.

$\frac{3}{4}$ ← numerator
← denominator

octagon
A 2D shape with eight sides.

regular octagon   irregular octagon

odd number
A number that is not divisible by two.

e.g. 1, 3, 5, 7, 9, 11

parallel lines
Two or more lines that never meet and are exactly the same distance apart along their entire length.

Dictionary

parallelogram

A four-sided 2D shape which has two pairs of opposite sides that are parallel and of equal length.

pattern

A series of shapes, letters, numbers or objects arranged in a recurring sequence.

e.g. 4, 14, 24, 34 ...

□ ○ □ ○ □ ○

pentagon

A 2D shape with five sides.

regular pentagon   irregular pentagon

perimeter

The distance around the edges of a shape.

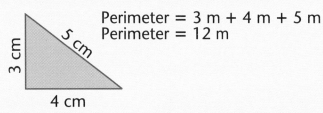

Perimeter = 3 m + 4 m + 5 m
Perimeter = 12 m

3 cm   5 cm   4 cm

perpendicular

A straight vertical line forming a right angle with the horizontal.

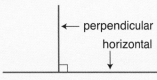

← perpendicular

horizontal

perpendicular lines

Lines that intersect at right angles.

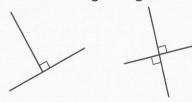

picture graph

A graph that uses symbols to represent quantities.

Animals of our class

horse   cat   chook   sheep   goat

place value

The value of a digit determined by its place in the number.

346 → 4 = 4 tens

495 → 4 = 4 hundreds

704 → 4 = 4 ones

plan

A diagram from above, showing the position of objects.

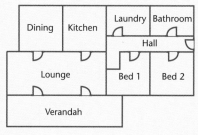

Dining | Kitchen | Laundry | Bathroom

Hall

Lounge | Bed 1 | Bed 2

Verandah

plane shape

A 2D shape.

Example: 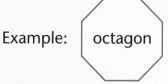 octagon

p.m. (*post meridiem*)

After midday. Any time between noon and midnight,
e.g. 7:30 p.m. is 7:30 in the evening.

polygon

A 2D shape with three or more angles and straight sides.

Example: 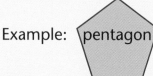 pentagon

prism

A 3D shape that has a pair of congruent parallel bases that are polygons, and rectangular side faces.

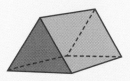

Dictionary

product
The answer achieved by multiplication.

$$9 \times 3 = 27$$

product

pyramid
A 3D shape that has only one base and all the other faces are triangular, meeting at a point (the apex).

quadrilateral
A 2D shape with four sides, e.g. a square, a rectangle, a rhombus and a parallelogram.

rhombus

quarter
One-fourth of a whole or a group.

 $= \dfrac{1}{4}$

 $\dfrac{1}{4}$ are girls

rectangle
A four-sided 2D shape with four right angles and two pairs of parallel sides that are also equal.

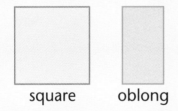

square   oblong

reflect (flip)
To turn a shape over so it appears like a mirror image of itself.

rhombus
A four-sided, 2D shape with all four sides equal. A rhombus has two pairs of parallel sides and its opposite angles are equal.

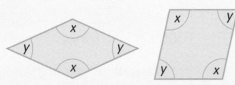

right angle
An angle of 90°.

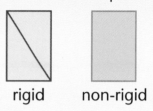

rigid
A fixed shape that cannot be pulled out of shape.

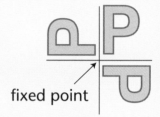

rigid   non-rigid

rotate (turn)
To turn an object around a fixed point.

fixed point

rounding
Changing an exact value to an estimated value of a more convenient size.

$69 \longrightarrow 70$ (rounded to the nearest 10)

$785 \longrightarrow 800$ (rounded to the nearest 100)

side view
The view we see when we look at an object from the side.

3D shape

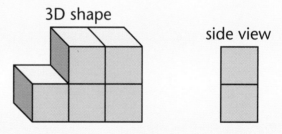

side view

skip counting
To count on by adding the same number each time.

4.. 8.. 12.. 16.. 20..
Skip counting by 4.

Dictionary

slide (translate)
Move a shape to a new position without turning it.

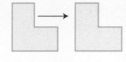

sphere
A perfectly round 3D shape, e.g. a ball.

square
A 2D shape with four equal sides and four right angles. A square is also a rectangle.

square centimetre (cm²)
A unit for measuring area.

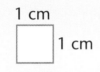

1 cm × 1 cm = 1 cm²

square kilometre (km²)
A unit for measuring area.

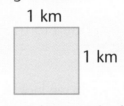

1 km × 1 km = 1 km²

square metre (m²)
A unit for measuring area.

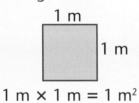

1 m × 1 m = 1 m²

subtraction (−)
The operaton which removes part of a group, and finds the difference.

9 − 5 = 4
↑
difference

sum
The answer of addition. The total.

symmetry
A shape has line symmetry if both parts match when it is folded along a line.

tally
To keep count by placing a stroke to represent each item. The fifth stroke crosses the four preceding strokes each time.

̶H̶H̶ ̶H̶H̶ ̶H̶H̶ ̶H̶H̶ || = 22

tessellation
A pattern formed by the repetition of shapes so that they fit together without gaps.

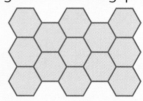

three-dimensional (3D)
A description of solid objects having three dimensions: length, width and height.

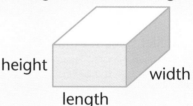

time
60 seconds = 1 minute
60 minutes = 1 hour
24 hours = 1 day
365 days = 1 year
366 days = a leap year

timeline
A line representing a span of time.

Born	School	High school		Work	Married
1985	1990	1995	2000	2005	2010

tonne
A unit of mass

1000 kg = 1 tonne

top view
The view we see when we look at an object from the top.

3D shape

top view

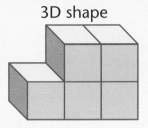
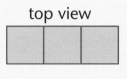

translate (slide)
Move a shape to a new position without turning it.

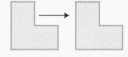

trapezium
A four-sided, 2D shape with only one pair of parallel sides.

triangle
A 2D shape with three sides and three angles.

turn (rotate)
To move an object around a fixed point.

twelve-hour time
Time shown by traditional clocks and watches, divided into 12 hours.

twenty-four hour time
Time divided into 24-hour time intervals numbered from 1 to 24, so as to distinguish between times in the morning and times in the afternoon.

two-dimensional (2D)
Plane shapes have only two dimensions: length and width.

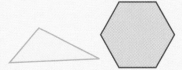

vertex
The point where two or more lines meet to form an angle.

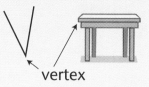

vertex

vertical
At right angles to the horizontal.

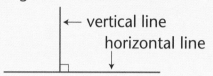

← vertical line
horizontal line

volume
The amount of space a 3D shape occupies.

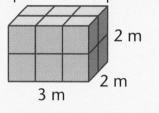

2 m

2 m

3 m

$$volume = 2 \times 3 \times 2 = 12 \text{ m}^3$$

whole numbers
The counting numbers from one to infinity.

$1, 2, 3, 4, \longrightarrow$

width
The lesser of the two dimensions of a shape.

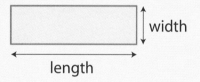

width

length

Answers

Unit 1

1 a 51 b 92 c 84 d 94

2 a 153 b 273 c 284 d 363 e 473 f 352

3 a 130 1300 e 570 5700 12000
  b 90 900 f 860 8600 5000
  c 110 1100 g 740 7400 2000
  d 210 2100

4

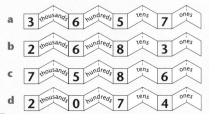

a 3657	3	6	5	7
b 2683	2	6	8	3
c 7586	7	5	8	6
d 2074	2	0	7	4

a 3 thousands 6 hundreds 5 tens 7 ones
b 2 thousands 6 hundreds 8 tens 3 ones
c 7 thousands 5 hundreds 8 tens 6 ones
d 2 thousands 0 hundreds 7 tens 4 ones

5 a hundreds d tens g ones
  b hundreds e thousands h hundreds
  c tens f thousands i ones

6 a 563 564 565 d 5461 5462 5463
  b 386 387 388 e 6998 6999 7000
  c 399 400 401 f 3000 3001 3002

7 a 46, 56, 247, 474
  b 323, 357, 531, 784
  c 2374, 2743, 3724, 7423
  d 2671, 2701, 2761, 3017
  e 3806, 6380, 6803, 8603

8

Cylinders	f	j	
Cones	b	c	g
Spheres	a	h	
Prisms	d e	k	l
Pyramids	i	m	

9 Hands on. **10** Hands on. (It has 2 bases.)

11 a 15 cm c 6 cm e 11 cm
  b 7 cm d 9 cm f 3 cm

12 Hands on. **13** Hands on.

14 a 50 cm c 25 cm e 1 m g 75 cm
  b 50 cm d 25 cm f 100 cm h 75 cm

15 a 10 cm c 12 cm
  b 8 cm d 11 cm

Unit 2

1 a 62 b 22 c 25 d 51 e 18
  f 22 g 29 h 27 i 29 j 32

2 a 2, 20, 200 e 4, 40, 400
  b 4, 40, 400 f 2, 20, 200
  c 3, 30, 300 g 1, 10, 100
  d 5, 50, 500

3 a 48 d 138 g 593 j 706
  b 42 e 249 h 485 k 577
  c 68 f 536 i 219 l 291

4 Some solutions.
  a b c d

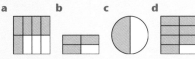

5 Possible solutions.
  a
  b
  c
  d

6 Possible solutions:
  a ⊙⊙⊙⊙ b ▦▦▦ c ▦▦
    ⊙⊙⊙⊙

7 a $\frac{1}{2}$ b $\frac{1}{4}\frac{2}{4}\frac{3}{4}$ c $\frac{1}{8}\frac{2}{8}\frac{3}{8}\frac{4}{8}\frac{5}{8}\frac{6}{8}\frac{7}{8}$

8 a $\frac{1}{8}\frac{3}{8}\frac{5}{8}\frac{7}{8}$ b $\frac{1}{4}\frac{1}{4}\frac{2}{4}\frac{3}{4}$

9 a Jimmy e Nick and Aimee
  b Angel f Simone
  c Angel g Nick and Aimee
  d 3 h Alex

10 Hands on.

11 a 6 cm b 8 cm c 10 cm d 4 cm

12–14 Hands on.

Unit 3

1 a 9, 12, 18, 15, 21, 6, 24, 30, 27
  b 12, 16, 24, 20, 28, 8, 32, 40, 36
  c 18, 24, 36, 30, 42, 12, 48, 60, 54

2 a

The Younis family	
Item	Cost
10 kg of potatoes	$20
3 kg of mushrooms	$18
2 cabbages	$ 6
6 kg of beans	$12
4 kg of tomatoes	$16
Total	$72

b

The Walters family	
Item	Cost
5 kg of potatoes	$10
2 kg of mushrooms	$12
3 cabbages	$ 9
7 kg of beans	$14
3 kg of tomatoes	$12
Total	$57

3 Hands on.

4 a 100, 100 c 120, 120 e 110, 110
  b 110, 110 d 120, 120 f 118, 118

5 a 180, 180 d 240, 240 g 528, 528
  b 200, 200 e 170, 170 h Yes
  c 300, 300 f 260, 260

6 a 135 + 125 = 260
    125 + 135 = 260
  b 6 × 20 = 120
    20 × 6 = 120

7 a C9 c J6 e J10
  b A7 d G13 f J8

8 a Barney's Retreat c Railway Station
  b Last Post Bookshop

9 Hands on.

10 Hands on. **11** RSL

12 a 8 cm b 10 cm c 12 cm d 16 cm

13 Hands on.

14 a 12 cm c 12 cm e 14 cm
  b 12 cm d 18 cm f 18 cm

Unit 4

1 a 170, 170 c 308, 308 e 149, 149
  b 164, 164 d 356, 356

2 a 24, 24 c 40, 40 e 60, 60
  b 12, 12 d 60, 60

3 a 60 + 40 + 47 = 147
  b 75 + 25 + 23 = 123
  c 70 + 30 + 90 = 190
  d 180 + 20 + 38 = 238
  e 2 × 5 × 7 = 70
  f 4 × 5 × 7 = 140

4

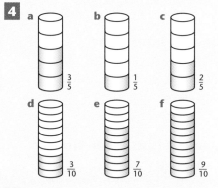

  a $\frac{3}{5}$ b $\frac{1}{5}$ c $\frac{2}{5}$
  d $\frac{3}{10}$ e $\frac{7}{10}$ f $\frac{9}{10}$

5 Possible solutions
  a b
  c

6 a $\frac{1}{2}$ c $\frac{1}{10}\frac{2}{10}\frac{3}{10}\frac{4}{10}\frac{5}{10}\frac{6}{10}\frac{7}{10}\frac{8}{10}\frac{9}{10}$
  b $\frac{1}{5}\frac{2}{5}\frac{3}{5}\frac{4}{5}$

7 a $\frac{3}{10}\frac{4}{10}\frac{5}{10}\frac{7}{10}$ d $\frac{1}{5}\frac{2}{5}\frac{3}{5}\frac{4}{5}$
  b $\frac{1}{10}\frac{3}{10}\frac{4}{10}\frac{3}{10}$ e $\frac{2}{5}\frac{1}{5}\frac{4}{5}\frac{5}{5}$
  c $\frac{3}{10}\frac{1}{2}\frac{8}{10}\frac{9}{10}$ f $\frac{1}{10}\frac{1}{5}\frac{1}{2}\frac{7}{10}$

8 Hands on.

9 a No b No c 32

Answers

10

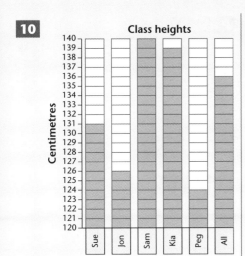

Class heights
Centimetres

11–12 Hands on.

Unit 5

1 **a** 412   **c** 762   **e** 741
  b 697   **d** 794   **f** 722

2
 a $623 + 174 = 797$
  $565 + 268 = 833$
 b $355 + 378 = 733$
  $268 + 586 = 854$

3
 a $285 + 15 + 30 = 330$
 b $496 + 4 + 30 = 530$
 c $295 + 5 + 20 = 320$
 d $588 + 12 + 20 = 620$
 e $392 + 8 + 30 = 430$
 f $490 + 10 + 53 = 553$
 g $297 + 3 + 80 = 380$
 h $584 + 16 + 60 = 660$
 i $675 + 25 + 20 = 720$

4
a	$2 \times 3 = 6$		$2 \times 30 = 60$
b	$2 \times 4 = 8$		$2 \times 40 = 80$
c	$3 \times 3 = 9$		$3 \times 30 = 90$
d	$3 \times 5 = 15$		$3 \times 50 = 150$
e	$3 \times 8 = 24$		$3 \times 80 = 240$
f	$5 \times 3 = 15$		$5 \times 30 = 150$
g	$5 \times 5 = 25$		$5 \times 50 = 250$
h	$5 \times 6 = 30$		$5 \times 60 = 300$
i	$4 \times 3 = 12$		$4 \times 30 = 120$
j	$4 \times 6 = 24$		$4 \times 60 = 240$
k	$4 \times 8 = 32$		$4 \times 80 = 320$

5 **a** 120   **c** 200   **e** 100
  b 60   **d** 120   **f** 200

6

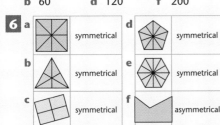

 a symmetrical   **d** symmetrical
 b symmetrical   **e** symmetrical
 c symmetrical   **f** asymmetrical

7

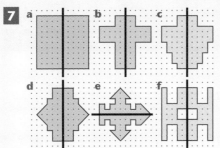

 a   **b**   **c**
 d   **e**   **f**

8 **a** 4 cm²   **b** 6 cm²   **c** 6 cm²
  d 10 cm²   **e** 10 cm²

9 **a** 16 cm²   **b** 20 cm²   **c** 15 cm²
  d 12 cm²   **e** 10 cm²   **f** 8 cm²
  g 14 cm²   **h** 28 cm²

Unit 6

1 **a** 21 **b** 35 **c** 63 **d** 70 **e** 77

2 **a** 42   **b** 21   **c** 49

3 **a** 21   **d** 35   **g** 48   **j** 24
  b 14   **e** 36   **h** 40
  c 28   **f** 9   **i** 42

4 Possible solutions:
  $2 \times 12,\ 3 \times 8,\ 4 \times 6,\ 24 \times 1$

5

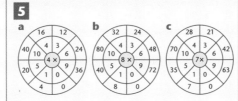

 a   **b**   **c**

6
 a $136 + 30 + 6 = 172$
 b $245 + 40 + 7 = 292$
 c $368 - 30 - 6 = 332$
 d $547 - 30 - 4 = 513$
 e $626 + 50 + 7 = 683$

7 **a** 459   **d** 427   **g** 461   **j** 942
  b 460   **e** 453   **h** 422
  c 436   **f** 426   **i** 899

8
 a $362 + 227 = 362...562...582 = 589$
 b $456 + 337 = 456...756...786...793 = 793$
 c $238 + 548 = 238...738...778...786 = 786$

9 **a** obtuse   **e** straight   **i** right
  b acute   **f** right   **j** acute
  c right   **g** obtuse   **k** straight
  d acute   **h** reflex   **l** obtuse

10 **a** c   **b** a   **c** b

11 **a** Mailbox = Acute
  b Door = Obtuse
  c Oven = Right

12–14 Hands on.

Unit 7

1
 a $50 + 12 = 62$
 b $70 + 11 = 81$
 c $60 + 13 = 73$
 d $70 + 13 = 83$
 e $170 + 13 = 183$
 f $150 + 14 = 164$
 g $170 + 12 = 182$
 h $270 + 14 = 284$
 i $360 + 12 = 372$

2
 a $400 + 70 + 8 = 478$
 b $600 + 80 + 7 = 687$
 c $600 + 70 + 9 = 679$
 d $500 + 80 + 9 = 589$
 e $800 + 60 + 7 = 867$
 f $800 + 70 + 9 = 879$
 g $400 + 80 + 6 = 486$

3 Hands on.

4 Some solutions:

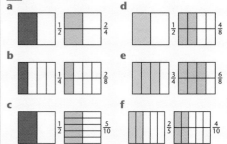

 a $\frac{1}{2}$ $\frac{2}{4}$ **d** $\frac{1}{2}$ $\frac{4}{8}$
 b $\frac{1}{4}$ $\frac{2}{8}$ **e** $\frac{3}{4}$ $\frac{6}{8}$
 c $\frac{1}{2}$ $\frac{5}{10}$ **f** $\frac{2}{5}$ $\frac{4}{10}$

5 **a** $\frac{2}{8} < \frac{1}{2}$ **d** $\frac{3}{4} = \frac{6}{8}$ **g** $\frac{3}{4} = \frac{6}{8}$
  b $\frac{1}{2} = \frac{5}{10}$ **e** $\frac{1}{2} > \frac{2}{5}$ **h** $\frac{4}{8} = \frac{1}{2}$
  c $\frac{7}{10} > \frac{1}{2}$ **f** $\frac{1}{5} < \frac{3}{10}$

6 **a** 135   **d** 25
  b Toyota   **e** She was fairly accurate.
  c BMW

7–10 Hands on.

11 **a** 16 cm²   **b** 15 cm²   **c** 24 cm²

Unit 8

1 **a** 749 **d** 859 **g** 289 **j** 47 **m** 102
  b 679 **e** 78 **h** 578 **k** 34
  c 578 **f** 448 **i** 51 **l** 80

2 Hands on.

3 **a** 589 km (or 647 km if Janice went to
    Melbourne via Canberra)
  b 240 km   **c** 359 km   **d** 1895 km

4 The missing numbers are:
  a 15, 18, 21 [30]   **c** 25, 30, 35 [50]
  b 20, 24, 28 [40]   **d** 30, 36, 42 [60]

5 **a** 36, 64 or 100 will fit.
  b 15, 21, 45 or 55 will fit.

Answers

6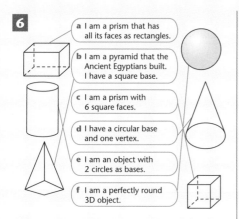
- **a** I am a prism that has all its faces as rectangles.
- **b** I am a pyramid that the Ancient Egyptians built. I have a square base.
- **c** I am a prism with 6 square faces.
- **d** I have a circular base and one vertex.
- **e** I am an object with 2 circles as bases.
- **f** I am a perfectly round 3D object.

7 Hands on.

8 Hands on. (Prisms have two bases and rectangular sides. Pyramids have one base and triangular sides.)

9–11 Hands on.

Unit 9

1
- **a** $160   **c** $96   **e** 132   **g** 124
- **b** $287   **d** 391   **f** 93   **h** $156

2

Number	Thousands	Hundreds	Tens	Ones
800		8	0	0
7 296	7	2	9	6
2 307	2	3	0	7
60			6	0
5 207	5	2	0	7
1 406	1	4	0	6
6 237	6	2	3	7

3
- **a** 8 507, 7 503, 5 073, 3 057
- **b** 3 658, 2 645, 2 500, 1 999
- **c** 8 436, 3 541, 2 907, 2 657
- **d** 5 234, 4 532, 3 524, 2 453
- **e** 2 745, 1 438, 837, 238

4
- **a** 653   **c** 432   **e** 7 432
- **b** 970   **d** 6 531

5
- **a** $4527 = \boxed{4000} + \boxed{500} + \boxed{20} + \boxed{7}$
- **b** $5436 = \boxed{5000} + \boxed{400} + \boxed{30} + \boxed{6}$
- **c** $6748 = \boxed{6000} + \boxed{700} + \boxed{40} + \boxed{8}$
- **d** $6740 = \boxed{6000} + \boxed{700} + \boxed{40}$
- **e** $8407 = \boxed{8000} + \boxed{400} + \boxed{7}$
- **f** $7987 = \boxed{7000} + \boxed{900} + \boxed{80} + \boxed{7}$
- **g** $8579 = \boxed{8000} + \boxed{500} + \boxed{70} + \boxed{9}$

6
- **a** North   **e** South   **i** East
- **b** East   **f** North   **j** North
- **c** North   **g** South   **k** Hands on.
- **d** South   **h** East

7 **a–g**
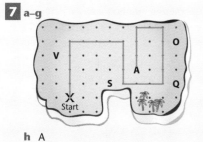
h A

8–15 Hands on.

DIAGNOSTIC REVIEW 1

Part 1
a

Number	Thousands	Hundreds	Tens	Ones
4 326	4	3	2	6
5 279	5	2	7	9
6 380	6	3	8	0
4 206	4	2	0	6
1 702	1	7	0	2

b $\boxed{2326}$ **c** 1 357, 3 571, 7 537

Part 2
- **a** 90   **b** 30   **c** 290   **d** 70
- **e** $\boxed{600} + \boxed{70} + \boxed{7} = \boxed{677}$
- **f** $\boxed{700} + \boxed{40} + \boxed{15} = \boxed{755}$
- **g** $\boxed{800} + \boxed{70} + \boxed{12} = \boxed{882}$
- **h** $\boxed{700} + \boxed{90} + \boxed{17} = \boxed{807}$

Part 3

×	2	3	4	5	6	7
a 2	4	6	8	10	12	14
b 4	8	12	16	20	24	28
c 7	14	21	28	35	42	49
d 5	10	15	20	25	30	35

- **e** 80 **f** 100 **g** 160 **h** 120 **i** 200 **j** 280
- **k** 50 × 3 = 150, 3 × 50 = 150
- **l** 60 × 4 = 240, 4 × 60 = 240
- **m** 42 × 5 = 210, 5 × 42 = 210
- **n** 80   **o** 140

Part 4
- **a** $\frac{1}{4}\frac{2}{4}\frac{3}{4}$ **b** $\frac{1}{5}\frac{2}{5}\frac{3}{5}\frac{4}{5}$ **c** $\frac{1}{4}$ **d** $\frac{3}{4}$

Part 5
Hands on.

Part 6
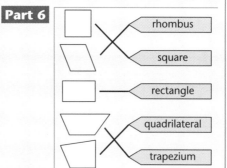
- rhombus
- square
- rectangle
- quadrilateral
- trapezium

Part 7   **a** East   **b** South
Part 8   **a** 7 cm   **b** 12 cm   **c** 14 cm
Part 9   **a** 8 cm² **b** 9 cm²

Part 10
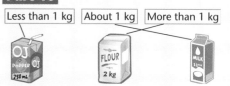
Less than 1 kg   About 1 kg   More than 1 kg

Part 11
- **a** Ryan   **b** Vicki   **c** Sue   **d** John

Unit 10

1
- **a** 66 **d** 88 **g** 72 **j** 52 **m** 56
- **b** 55 **e** 48 **h** 84 **k** 65 **n** 70
- **c** 77 **f** 60 **l** 96 **i** 78 **o** 84

2
- **a** 24 **d** 48 **g** 60 **j** 88
- **b** 32 **e** 56 **h** 64 **k** 100
- **c** 40 **f** 52 **i** 80 **l** 120

3
- **a** 40 **d** 60 **g** 100 **j** 30 **m** 48
- **b** 45 **e** 70 **h** 24 **k** 42 **n** 60
- **c** 50 **f** 75 **i** 36 **l** 54 **o** 72

4 **a** $1\frac{1}{4}$ **b** $2\frac{1}{2}$ **c** $2\frac{1}{4}$ **d** $1\frac{3}{10}$ **e** $3\frac{3}{8}$

5 Hands on. Some solutions below.
- **a** 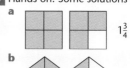 $1\frac{3}{4}$
- **b** $1\frac{3}{5}$
- **c** $2\frac{7}{8}$
- **d** $1\frac{4}{5}$
- **e** $2\frac{3}{8}$
- **f** $1\frac{3}{10}$

6

Tessellating	a	b	c	f
Non-tessellating	d	e		

7 Hands on.   **8** Hands on.

9
- **a** 4   **c** 1   **e** 2   **g** 5
- **b** 3   **d** 7   **f** 6

10 Hands on.   **11** Hands on.

Unit 11

1
- **a** 6 **d** 12 **g** 22 **j** 19
- **b** 8 **e** 15 **h** 33 **k** 23
- **c** 10 **f** 18 **i** 24 **l** 43

2
- **a** 3 **d** 7 **g** 9 **j** 2
- **b** 5 **e** 8 **h** 12 **k** 6
- **c** 4 **f** 10 **i** 11 **l** 13

3 **a** $20   **b** 216 km

Answers

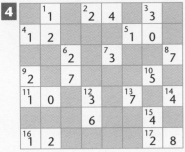

4

Across
2 4 × 6 = 24	11 50 ÷ 5 = 10
4 24 ÷ 2 = 12	13 28 ÷ 4 = 7
5 40 ÷ 4 = 10	14 24 ÷ 6 = 4
7 15 ÷ 5 = 3	16 36 ÷ 3 = 12
8 21 ÷ 3 = 7	17 7 × 4 = 28
10 25 ÷ 5 = 5	

Down
1 3 × 4 = 12	
3 3 × 10 = 30	
6 9 × 3 = 27	
9 7 × 3 = 21	
12 6 × 6 = 36	
15 7 × 6 = 42	

5
a 8	d 32	g 56	j 80
b 16	e 40	h 64	
c 24	f 48	i 72	

6
a 24	d 40	g 88	i 104
b 32	e 64	h 120	j 112
c 56	f 72		

7 Hands on. (Example: multiply by 4 then multiply by 2 or vice versa.)

8
1 42	5 49	9 56	13 56
2 56	6 72	10 80	14 40
3 40	7 28	11 40	15 40
4 28	8 56	12 56	16 63

1	2	3	4	5	6	7	8	9	10	11	12	13	14	15	16
G	R	E	A	T	B	A	R	R	I	E	R	R	E	E	F

9
Greece	Italy	Vietnam	Australia	Lebanon
6	5	7	8	4

10

Greece	
Italy	
Vietnam	
Australia	
Lebanon	

11–12 Hands on.

13
a 1 minute	f 1 year
b 1 hour	g 1 year
c 1 day	h 1 year
d 1 week	i 1 leap year
e 2 weeks	

14
a minutes/hours	f hours
b minutes/hours	g minutes
c seconds/minutes	h months
d weeks	i minutes
e seconds	j years

15
a <	c <	e >	g <	i >
b >	d <	f <	h >	j <

16

1 min 18 sec	1 min 19 sec	1 min 20 sec	1 min 12 sec	1 min 56 sec
second	third	fourth	first	fifth

Unit 12

1
a 9	c 27	e 45	g 63	i 81
b 18	d 36	f 54	h 72	j 90

2, 4

1	2	3	4	5	6	7	8	9	10
11	12	13	14	15	16	17	18	19	20
21	22	23	24	25	26	27	28	29	30
31	32	33	34	35	36	37	38	39	40
41	42	43	44	45	46	47	48	49	50
51	52	53	54	55	56	57	58	59	60
61	62	63	64	65	66	67	68	69	70
71	72	73	74	75	76	77	78	79	80
81	82	83	84	85	86	87	88	89	90

3 Because it makes a pattern.

5 Hands on (yes).

6 Hands on. (3 times tables are part of 9 times tables. Both follow a diagonal pattern.)

7

×	2	4	5	1	3	7	6	9	10	8
9	18	36	45	9	27	63	54	81	90	72
10	20	40	50	10	30	70	60	90	100	80
6	12	24	30	6	18	42	36	54	60	48
7	14	28	35	7	21	49	42	63	70	56
8	16	32	40	8	24	56	48	72	80	64

8
a 4 × 9 = 36	g 3 × 9 = 27
b 9 × 6 = 54	h 4 × 7 = 28
c 7 × 9 = 63	i 8 × 6 = 48
d 5 × 9 = 45	j 7 × 7 = 49
e 9 × 9 = 81	k 8 × 8 = 64
f 8 × 9 = 72	l 7 × 8 = 56

9 a, b, c — number lines

10 a, b — number lines

11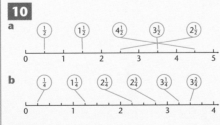

12–13 Hands on.

14 Hands on. (Prism with two hexagonal bases and six rectangular sides.)

15–16 Hands on.

17 a 10   b 5   c 2   d 4   e 20

Unit 13

1
a	347 + 40 = 387 subtract 2 = 385
b	455 + 20 = 475 subtract 1 = 474
c	563 + 30 = 593 subtract 3 = 590
d	245 + 30 = 275 subtract 1 = 274
e	373 + 40 = 413 subtract 2 = 411
f	166 + 40 = 206 subtract 1 = 205
g	276 + 50 = 326 subtract 2 = 324

2
a	234 + 50 = 284
b	361 + 60 = 421
c	540 + 690 = 1 230
d	593 + 260 = 853
e	645 + 300 = 945

3
a 666	c 390	e 589	g 681
b 296	d 305	f 699	h 781

4 Hands on (e.g. 150 + 34 = 184).

5 a 200 b 300 c 400 d 500 e 300

6
a 200	d 300	g 400	j 500
b 400	e 3 300	h 4 600	k 3 500
c 200	f 3 500	i 7 500	l 9 200

7
a	500 + 100 ≈ 600
b	1000 − 500 ≈ 500
c	400 + 100 ≈ 500
d	600 + 400 ≈ 1 000
e	900 − 500 ≈ 400
f	400 + 200 ≈ 600
g	700 + 200 ≈ 900
h	700 − 500 ≈ 200
i	600 + 200 ≈ 800

8

	Strategy	Approx Answer
a	78 tens − 25 tens = 53 tens	about 530
b	56 tens − 34 tens = 22 tens	about 220
c	68 tens − 47 tens = 21 tens	about 210
d	97 tens − 54 tens = 43 tens	about 430
e	85 tens − 24 tens = 61 tens	about 610
f	76 tens − 44 tens = 32 tens	about 320

9
a horizontal	e vertical
b vertical	f oblique
c oblique	g horizontal
d oblique	h vertical

10
a true	c true	e true
b true	d true	f false

11 Hands on.

12
a true	c true	e true
b false	d false	f true

13 green, pink, orange, yellow, red, black

14
a 4 red	c 2 blue	e 3 pink
b 7 green	d 9 yellow	

15 a yellow   b blue   c green

Answers

Unit 14

1
a 7868 c 5585 e 6689
b 7879 d 9999

2
a 7782 c 7683 e 6982
b 7962 d 9686

3
a 7815 c 7893 e 8723
b 3949 d 5822

4
a 324 b 924 kg c 2150 km
d 3864

5
a 3 d 6 g 6 j 6 m 5
b 4 e 4 h 3 k 6 n 10
c 7 f 7 i 8 l 7 o 9

6
a 6 d 8 g 5 j 4
b 5 e 4 h 5 k 5
c 6 f 9 i 3 l 7

7
a 4 × 8 = 32, or 8 × 4 = 32
b 5 × 7 = 35, or 7 × 5 = 35
c 6 × 7 = 42, or 7 × 6 = 42

8
a 12 b 8 c 6 d 4

9 Hands on.

10
a 8   g Rock
b 3   h Ego or Guitar
c 2   i no
d no   j Hotville
e Hotville k Cool and Rap
f 2

11
a East b West c South
d South e North f West

12 Hands on.

13 a Hands on. b Janice

Unit 15

1
a 60 + 54 = 114 f 150 + 45 = 195
b 50 + 40 = 90 g 80 + 32 = 112
c 60 + 3 = 63 h 400 + 16 = 416
d 150 + 15 = 165 i 150 + 24 = 174
e 280 + 7 = 287

2
a 120 d 150 g 120
b 100 e 280 h 400
c 60 f 200 i 180

3 a 76 km b $240

4
a $\frac{33}{100}$ / 0.33 d $\frac{88}{100}$ / 0.88
b $\frac{54}{100}$ / 0.54 e $\frac{40}{100}$ / 0.40
c $\frac{62}{100}$ / 0.62

5
a Shade any 35 squares.
b Shade any 38 squares.
c Shade any 28 squares.
d Shade any 72 squares.
e Shade any 57 squares.

6 Possible solutions

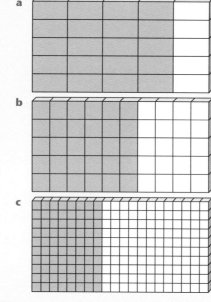
a
b
c

7
a
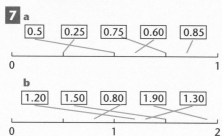
0.5 0.25 0.75 0.60 0.85
0 _____ 1
b
1.20 1.50 0.80 1.90 1.30
0 _____ 1 _____ 2

8
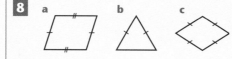
a b c

9
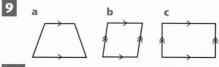
a b c

10

a b c

11
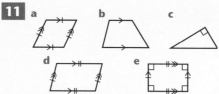
a b c
d e

12
a Colour 100 mL on beaker
b Colour 200 mL on beaker
c Colour 300 mL on beaker
d Colour 400 mL on beaker
e Colour 500 mL on beaker
f Colour 1 000 mL on beaker

13–14 Hands on.

Unit 16

1
a 5 243 c 1 432 e 3 211
b 2 635 d 3 553

2
a 5 046 c 2 219 e 4 715
b 5 428 d 1 116

3
a 5 653 c 4 226 e 825
b 3 436 d 5 948

4
a $6 055 c 3 197
b 6 095 m² d 1 228

5
a 150, 160 f 1 000, 1 200
b 280, 270 g 590, 490
c 350, 390 h 1 520, 2 020
d 190, 140 i 1 650, 1 750
e 460, 360 j 4 260, 5 260

6
a 30, 40, 50, 60, 70
b 200, 300, 400, 500, 600
c 166, 176, 186, 196, 206
d 284, 274, 264, 254, 244
e 986, 886, 786, 686, 586
f 3 313, 3 513, 3 713, 3 913, 4 113
g 5 428, 5 378, 5 328, 5 278, 5 228
h 1 526, 2 526, 3 526, 4 526, 5 526
i 1 238, 1 738, 2 238, 2 738, 3 238
j 9 025, 8 925, 8 825, 8 725, 8 625

7 a circle 86 b circle 789 c circle 1 820

8

Plastic Cans Paper Glass Food

9 Hands on. A possible solution:

Name	Colours	Name	Sport
red pink yellow purple green		hockey softball football cricket soccer	

Name	Subjects	Name	Planets
Maths English		Earth Saturn Mars Venus Jupiter Uranus	

10–12 Hands on.

13
a 1 500 g f 2 300 g
b 6 500 g g 2 700 g
c 2 500 g h 3 500 g
d 3 700 g i 1 800 g
e 2 600 g j 2 100 g

Unit 17

1 In order 0.1, 0.2, 0.3, 0.4, 0.5, 0.6, 0.7, 0.8, 0.9, 1

2
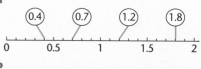
a
0.4 0.7 1.2 1.8
0   0.5   1   1.5   2
b
2.2 2.6 2.9 3.1 3.4 3.9
2   3   4

Answers

3

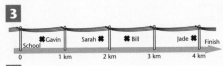

4
a 2, 2.1, 2.2, 2.3, 2.4, 2.5, 2.6, 2.7, 2.8, 2.9
b 5, 5.5, 6, 6.5, 7, 7.5, 8, 8.5, 9, 9.5
c 10, 9.8, 9.6, 9.4, 9.2, 9, 8.8, 8.6, 8.4, 8.2

5
a

Hours	1	2	3	4	5	6
Pay	5	10	15	20	25	30

Rule ×5

b

Minutes	1	2	3	4	5	6
Litres	3	6	9	12	15	18

Rule ×3

c

Bags	1	2	3	4	5	6
Kilos	4	8	12	16	20	24

Rule ×4

d

Hours	1	2	3	4	5	6
Kms	6	12	18	24	30	36

Rule ×6

e

Boxes	1	2	3	4	5	6
Pages	7	14	21	28	35	42

Rule ×7

f

Booklets	1	2	3	4	5	6
Pages	8	16	24	32	40	48

Rule ×8

g

Hours	1	2	3	4	5	6
Tom's Pay	9	18	27	36	45	54
Sam's Pay	10	20	30	40	50	60

Tom ×9, Sam ×10

6

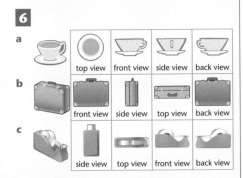

a top view front view side view back view
b front view side view top view back view
c side view top view front view back view

7

	Front view	Side view	Top view	Back view
a				
b				

8
a 25 past 3
b 15 past 9
c 20 to 2
d 10 to 4
e Half past 7 or 30 past 7
f 25 past 5
g 20 to 10
h 10 past 4
i 25 to 9
j 5 to 8
k 5 past 12

9 a 5   b 30   c 60

10 a 1   b 3   c 12

11 60

12 a 5   b 15   c 60

Unit 18

1 a 6962   c 8883   e 7832
       b 7890   d 8964

2 a 9483   c 5824   e 5975
       b 1449   d 7013

3 a 5954   c 7129   e 5143
       b 8542   d 7312

4 a $17.50   b $23.50

5 Hands on.

6 a 21, 34, 55, 89, 144, 233, 377
       b Add the previous number to the very last number to get the next number in line.

7 a The trains are 20 minutes apart.
       b The train takes 10 minutes to make it to the next station.
       c Hands on.

8 a 145, 155, 165, 175, 185, 195
       b 11.1, 11.2, 11.3, 11.4, 11.5, 11.6
       c 750, 650, 550, 450, 350, 250
       d 4, 40, 400, 4 000, 40 000, 400 000

9

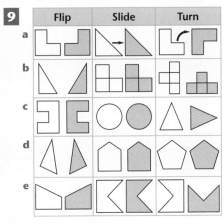

	Flip	Slide	Turn
a			
b			
c			
d			
e			

10

11
a Most likely **orange**
   Least likely **green**
b Most likely **green**
   Least likely **red**
c Most likely **yellow**
   Least likely **red**
d Most likely **pink**
   Least likely **blue**
e Hands on.   f Hands on.

12–13 Hands on.

DIAGNOSTIC REVIEW 2

Part 1

a 5 891   b 8 934   c 5 936   d 2 078
e 9 000   f 6 000   g 2 000

Part 2

a 28   b 56   c 160   d 72   e 72
f 54   g 80   h 70   i 90   j 45
k 72   l 128   m 96   n 160
o 8, 16, 24, 32, 40, 48, 56
p 4, 8, 12, 16, 20, 24, 28
q 9, 18, 27, 36, 45, 54, 63

Part 3

a 3   b 4   c 5   d 5   e 5   f 8

Part 4

a $0, \frac{1}{4}, \frac{2}{4}, \frac{3}{4}, 1, 1\frac{1}{4}, 1\frac{2}{4}, 1\frac{3}{4}, 2$

b $\frac{3}{5}, \frac{4}{5}, 1, 1\frac{1}{5}, 1\frac{2}{5}, 1\frac{3}{5}, 1\frac{4}{5}, 2$

c $2\frac{1}{4}, 2\frac{2}{4}, 2\frac{3}{4}, 3, 3\frac{1}{4}, 3\frac{2}{4}, 3\frac{3}{4}$

Part 5

Possible solutions:

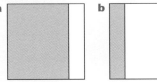

a   b

c 0.4, 0.6, 0.8, 0.9

Part 6

a 5, 10, 15, 20, 25
b 12, 24, 36, 48, 60

Part 7 a   b

Part 8

Part 9

Top	Front	Side	Back

Part 10

a Colour 300 mL   b Colour 700 mL

Part 11

Hands on.

Part 12

7 cm²

Part 13

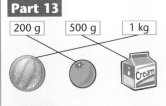

200 g   500 g   1 kg

Part 14

a Green   b Red   c No, less.

Answers

Unit 19

1 a accurate   c wrong (1 609)
b wrong (9 248)   d accurate

2 She left 7 in the tens column instead of placing 6 there after she traded a ten into the ones column.

3 a $6 707   d $6 927
b 4 245 mL   e 7 284 kg
c 8 264 tickets

4 a Colour in 5 $10 \times \frac{1}{2} = 5$
b Colour in 9 $18 \times \frac{1}{2} = 9$
c Colour in 6 $12 \times \frac{1}{2} = 6$

5 a Draw 2 more $4 \times 1\frac{1}{2} = 6$
b Draw 3 more $6 \times 1\frac{1}{2} = 9$
c Draw 4 more $8 \times 1\frac{1}{2} = 12$

6 a 12   b 36

7 a 60, 55, 50, 45, 40, 35, 30, 25
b 15, 18, 21, 24, 27, 30, 33, 36

8 a 90 minutes   b 1 500 g

9 a slide   c flip   e turn
b turn   d turn   f slide

10
a
b

11 Hands on.

12

13–16 Hands on.

Unit 20

1 a 3r1   c 4r1   e 3r1   g 4r2   i 6r2
b 3r1   d 2r2   f 5r1   h 5r2   j 5r1

2 Possible solutions:
a 9r1: 2 groups of 9, 1 group of 10
b 7r2: 5 groups of 7, 1 group of 9
c 9r3: 4 groups of 9, 1 group of 12
d 5r3: 3 groups of 5, 1 group of 8
e 5r5: 4 groups of 8, 1 group of 13
f 8r2: 5 groups of 8, 1 group of 10

3–4 Hands on.

5 0.69 m, 0.72 m, 0.89 m, 0.93 m, 1.04 m, 1.12 m

6–9 Hands on.

10 a 6 cm   c $8\frac{1}{2}$ cm   e 11 cm
b 3 cm   d 3 cm   f 4 cm

11 a Triangle   b 12 cm

12 a Rectangle   b 20 cm

13

14 a 1 542 cm—4th   d 1 995 cm—1st
b 1 238 cm—5th   e 1 808 cm—2nd
c 1 627 cm—3rd

Unit 21

1 a 4r1   e 6r1   i 3r1   m 3r1
b 4r1   f 5r2   j 3r1   n 3r3
c 4r1   g 7r1   k 4r2   o 4r3
d 4r1   h 5r1   l 4r1   p 7r1

2

	Division fact	My solution
a	27 ÷ 3 = 9	2 teams of 9 and 1 team of 11.
b	16 ÷ 4 = 4	3 groups of 4 and 1 group of 7.
c	40 ÷ 8 = 5	4 groups of 8 and 1 group of 10.
d	35 ÷ 5 = 7	4 groups of 7 and 1 group of 10.
e	30 ÷ 6 = 5	5 groups of 5 and 1 group of 6.
f	45 ÷ 9 = 5	8 groups of 5 and 1 group of 6.
g	64 ÷ 8 = 8	8 groups of 8.

3 a 9 cards   c $9 each
b 6 children   d 12 marbles

4

Con.	$\frac{1}{10}$s	$\frac{1}{100}$s	Dec.	Con.	$\frac{1}{10}$s	$\frac{1}{100}$s	Dec.
a	$\frac{3}{10}$	$\frac{30}{100}$	0.30	e	$\frac{4}{10}$	$\frac{40}{100}$	0.40
b	$\frac{2}{10}$	$\frac{20}{100}$	0.20	f	$\frac{7}{10}$	$\frac{70}{100}$	0.70
c	$\frac{6}{10}$	$\frac{60}{100}$	0.60	g	$\frac{5}{10}$	$\frac{50}{100}$	0.50
d	$\frac{8}{10}$	$\frac{80}{100}$	0.80	h	$\frac{9}{10}$	$\frac{90}{100}$	0.90

5 a <   d >   g >   j <   m <
b >   e >   h =   k <   n >
c >   f =   i >   l =   o <

6 a Shade 10 red   d Shade 30 blue
b Shade 30 yellow   e Shade 7 orange
c Shade 20 green   f Leave 3 white

7 a equilateral   d isosceles
b isosceles   e scalene
c equilateral   f scalene

8 Colour A, B, D

9 Hands on.

10

	Digital time	Words	Meaning
a	2:45 a.m.	Two forty-five a.m.	45 minutes past 2 in the morning
b	9:15 p.m.	Nine fifteen p.m.	15 minutes past 9 at night
c	6:29 p.m.	Six twenty-nine p.m.	29 minutes past 6 at night
d	8:48 a.m.	Eight forty-eight a.m.	12 minutes to 9 in the morning
e	9:21 p.m.	Nine twenty-one p.m.	21 minutes past 9 at night
f	11:35 a.m.	Eleven thirty-five a.m.	25 minutes to 12 in the morning

11 a 12 minutes   d 55 minutes
b 25 minutes   e 31 minutes
c 35 minutes

12
a A quarter past 6   b Half past 3   c A quarter to 5   d A quarter past 9
06:15   03:30   04:45   09:15

Unit 22

1 a 16   d 14   g 15   j 12   m 12
b 14   e 15   h 25   k 17   n 13
c 26   f 13   i 16   l 16   o 14

2 a 19 − 4 = 15, 3 × 5 = 15 correct
b 24 − 4 = 20, 4 × 5 = 20 correct
c 27 − 2 = 25, 6 × 4 = 24 incorrect
d 19 − 3 = 16, 4 × 4 = 16 correct
e 26 − 2 = 24, 3 × 8 = 24 correct
f 48 − 3 = 45, 5 × 9 = 45 correct
g 95 − 5 = 90, 9 × 10 = 90 correct

3 Discuss dealing with remainders.
a 7r2   b 8r3   c 6   d 19r3

4 a 0.43   c 0.22   e 0.54   g 0.90
b 0.55   d 0.37   f 0.14   h 0.07

5 a 0.3 m, 0.4 m, 0.6 m, 0.7 m
b 0.3 kg, 0.6 kg, 0.9 kg, 1.0 kg
c 0.16 m, 0.27 m, 0.63 m, 0.94 m
d $4.32, $4.36, $4.37, $4.39
e $3.26, $3.36, $3.67, $3.74
f 1.32 m, 2.14 m, 2.36 m, 3.26 m
g $24.93, $25.34, $42.61, $61.82

6–8 Hands on.

9

Model	Cubes per layer	No. of layers	Volume
A	3	1	3 cubes
B	2	2	4 cubes
C	3	2	6 cubes
D	4	2	8 cubes
E	6	2	12 cubes
F	6	3	18 cubes

10 a 12 cubes   b 24 cubes

Answers

Unit 23

1 a true b false c true d false e true f true

2 a 4 b 9 c 6 d 4

3 8 × 2 × 5 = 80
6 × 2 × 5 = 60
7 × 2 × 5 = 70
9 × 2 × 5 = 90

4 Possible answers:
a 2 × 5 × 6 = 60
b 2 × 5 × 10 = 100
c 2 × 5 × 14 = 140

5 a 1 b 1 c 2 d 2 e 3

6 a $2 b $3 c $3 d $4 e $2 f $2 g $9 h $5 i $3 j $8 k $3 l $2

7 a $12 b $20 c $12 d $17 e $28 f $19

8 a $5 b $12 c $15 d $50 e $404 f $185

9 a Yes b By rounding to the nearest whole dollar.

10 Hands on.

11 Expected answers: a–h yes
i right angles

12 Hands on.

13 a 20 kg b 45 kg c 25 kg d 32 kg

14 a 2 kg 900 g 2.9 kg
b 2 kg 500 g 2.5 kg
c 2 kg 200 g 2.2 kg
d 1 kg 500 g 1.5 kg

15 a Needle showing 1.3 kg
b Needle showing 2.4 kg
c Needle showing 2.9 kg
d Needle showing 3.3 kg

Unit 24

1 a 6333 b 2511 c 8332

2 a 5238 b 5117 c 2426 d 5319 e 3228

3 a 3327 b 5392 c 2274 d 4191 e 3571

4 a 752 b 3400 − 1150 = 2250 c 4152 − 1150 = 3002

5 a Longreach by 516 km
b Normanton by 1177 km
c Mackay by 231 km
d Normanton by 408 km

6 a 100 b 108 c 99 d 220 e 144 f 216 g 185 h 270 i 182 j 152 k 225 l 378 m 525 n 304 o 423

7 a $135 b $252

8 a square b rectangle c pentagon d circle

9

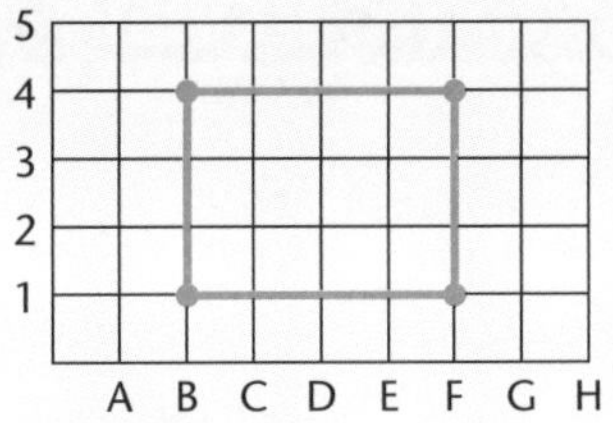

10 and 11

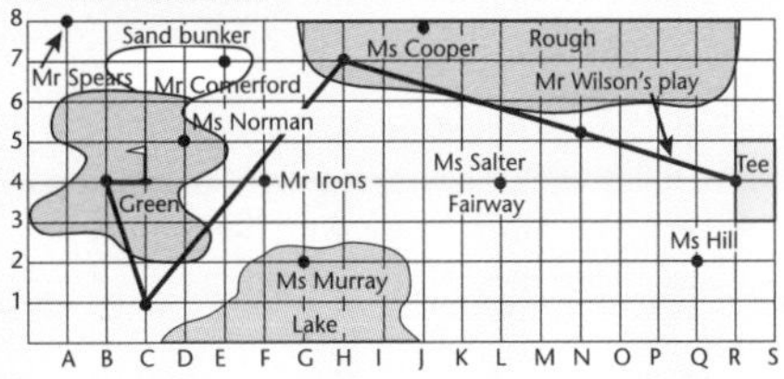

12 a blue b pink c white d yellow and green e red

13 a red = 1 b blue = 2 c green = 8 d yellow = 4

14 Each selection is an independent event and therefore the chance of any one marble being selected is equally likely.

Unit 25

1 a 69 b 52 c 68 d 81 e 112 f 104 g 140 h 180 i 162 j 114 k 150 l 108 m 138

2 a 100 b 63 c 128 d 168 e 126 f 336 g 132 h 90 i 120 j 108

3 a $144 b 216 cakes

4–5

1	2	3	4	5	6	7	8	9	10
11	12	13	14	15	16	17	18	19	20
21	22	23	24	25	26	27	28	29	30
31	32	33	34	35	36	37	38	39	40
41	42	43	44	45	46	47	48	49	50
51	52	53	54	55	56	57	58	59	60
61	62	63	64	65	66	67	68	69	70
71	72	73	74	75	76	77	78	79	80
81	82	83	84	85	86	87	88	89	90
91	92	93	94	95	96	97	98	99	100

6 Hands on. (It makes a pattern.)

7 Hands on. (Multiples of 5 or adding 5 from zero.)

8

1	2	3	4	5	6	7	8	9	10
11	12	13	14	15	16	17	18	19	20
21	22	23	24	25	26	27	28	29	30
31	32	33	34	35	36	37	38	39	40
41	42	43	44	45	46	47	48	49	50
51	52	53	54	55	56	57	58	59	60
61	62	63	64	65	66	67	68	69	70
71	72	73	74	75	76	77	78	79	80
81	82	83	84	85	86	87	88	89	90

9 9, 18, 27, 36, 45, 54, 63, 72, 81

10 She has created the 9's facts.

11–14 All hands on.

15 a 1500 millilitres 1 litre 500 mL
b 1400 millilitres 1 litre 400 mL
c 1700 millilitres 1 litre 700 mL

16 a 800 mL b 750 mL c 650 mL d 500 mL e 720 mL f 400 mL g 760 mL h 820 mL

17

	Millilitres	Litres
a	1800 mL	1.8 L
b	1600 mL	1.6 L
c	1900 mL	1.9 L
d	1250 mL	1.25 L
e	2300 mL	2.3 L
f	5400 mL	5.4 L
g	3450 mL	3.45 L
h	4500 mL	4.5 L

Unit 26

1 a 140 b 75 c 215 d 296 e 116 f 216 g 344 h 574 i 378 j 170 k 232 l 288 m 216 n 204 o 441

2 a 432 b 252 c $624 d 92 m e 48 km f 392 m

3 a $195 b $613 c $167 d $409 e $995 f $826 g $380

4 Discussion: Don't have to carry large amounts of money.

5 Discussion: Lost cards, computer errors.

6 a obtuse b acute c acute d acute e reflex f obtuse g acute h obtuse i reflex j acute k straight

7

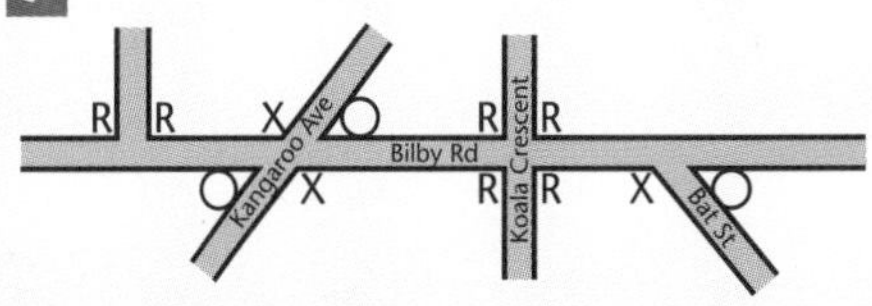

8 a Yes (50°)
b Hands on: The angle between the arms is the same even though the arms are different lengths.

9 a Wednesday b Thursday c Monday d Wednesday e Monday f Friday g Saturday h Tuesday

10 a 14 b 10 c 19 d 14 e 16

11 a 05/09/72 b 30/10/76 c 19/11/97 d 11/01/01

12 Hands on.

Answers

Unit 27

1
a 14r1 e 24r2 i 14r1 m 12r1
b 14r1 f 13r2 j 17r2 n 16r2
c 13r1 g 15r1 k 15r2 o 14r1
d 13r2 h 13r3 l 12r2

2
a 15 d 12 g 13 j 13 m 13
b 13 e 12 h 14 k 16 n 14
c 24 f 13 i 14 l 14 o 12

3
a 80 d 120 g 100 j 112
b 40 e 60 h 50 k 56
c 20 f 30 i 25 l 28

4 Hands on.

5
a ≠ c ≠ e ≠ g ≠
b ≠ d = f = h ≠

6
a 3 × 80 = 4 × 60
b 3 × 100 = 10 × 30
c 5 × 80 ≠ 7 × 60

7
a 5 × 3 × 2 > 2 × 6 × 2
b 3 × 6 × 5 > 100 − 18
c 4 × 7 × 3 > 100 − 84
d 5 × 9 × 2 < 2 × 5 × 10
e 3 × 2 × 7 > 2 × 8 × 2
f 48 + 9 + 12 < 5 × 7 × 2
g 63 − 48 − 10 < 100 ÷ 2 ÷ 2
h 50 + 40 + 30 = 10 × 2 × 6
i 85 − 40 + 15 < 10 × 3 × 5
j 75 − 30 + 22 < 17 × 2 × 2

8
a $1 000 + $600 + $500 > $2 000
b $800 + $800 + $350 < $2 000
c $500 × 5 > $2 000
d $200 × 7 + 500 < $2 000

9 Suggested answers:
a interval AB = 60 mm
b interval CD = 75 mm
c interval EF = 35 mm
d interval GH = 80 mm

10 Suggested answers:
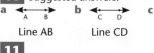
Line AB   Line CD   Line EF

11
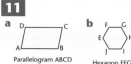
Parallelogram ABCD   Hexagon EFGHIJ   Pentagon KLMNO
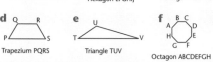
Trapezium PQRS   Triangle TUV   Octagon ABCDEFGH

12 Six possible combinations.
Ⓡ Ⓨ Ⓑ   Ⓑ Ⓡ Ⓨ
Ⓡ Ⓑ Ⓨ   Ⓨ Ⓡ Ⓑ
Ⓑ Ⓨ Ⓡ   Ⓨ Ⓑ Ⓡ

13 Nine combinations of colours.

Tops	white	white	white	pink	pink
Skirts	blue	red	green	blue	red

Tops	pink	gold	gold	gold
Skirts	green	blue	red	green

14
a any 3 squares red   c four or 5 red
b all 6 squares red   d none red

DIAGNOSTIC REVIEW 3

Part 1
a 5 881 d 6 393 f 3 636 h 6 214
b 7 938 e 1 886 g 7 696 i 8 179
c 4 789

Part 2
a 78   b 190   c 188
d 5 remainder 1
e 5 remainder 2
f 8 remainder 1
g 9 remainder 2
h 8   i 8   j 7   k 4r2
l 3r3 m 4r3 n 21 points

Part 3
a 40, 35, 30, 25, 20, 15
b 30, 60, 90, 120, 150, 180

Part 4
a 0.2, 0.3, 0.7   b 1.2, 1.3, 1.5
c 0.27, 0.35, 0.47   d 1.2, 1.27, 1.3
e 5   f 5   g 3
h 1   i $6   j $9
k $\frac{6}{10}$ $\frac{60}{100}$ 0.6   l $\frac{5}{10}$ $\frac{50}{100}$ 0.5
m $\frac{4}{10} < \frac{50}{100}$ n 0.6 = $\frac{6}{10}$
o $\frac{80}{100} < 0.90$ p $\frac{1}{2}$ = 0.50

Part 5 Possible answers
a   b c

Part 6 a   b

Part 7
On an equilateral triangle all sides and angles are equal.
On an isosceles triangle only 2 sides and 2 angles are equal.

Part 8
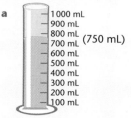

Part 9
a 10 mins c 30 mins e 35 mins
b 20 mins d 40 mins

Part 10 a
b 1 500 mL
c 2 375 mL
1000 mL
900 mL
800 mL (750 mL)
700 mL
600 mL
500 mL
400 mL
300 mL
200 mL
100 mL

Part 11
a T   b T   c T

Unit 28

1
a $2.00 d $2.05 g $2.10 j 10 cents
b $2.00 e $2.05 h $2.10
c $2.05 f $2.05 i $2.05

2
a $14.28, $14.30 c $15.37, $15.35
b $4.04, $4.05   d $14.93, $14.95

3
a $8.50   b $3.75

4 Hands on.

5
a 8.96 e 97.64 i 43.95 m 67.98
b 9.95 f 25.35 j 32.77 n 99.73
c 9.61 g 45.19 k 68.97 o 97.11
d 90.27 h 63.18 l 66.84

6
a 5.999 m   c 3.286 m
b 6.014 m   d 1.471 m

7
a 0.356 m   b 14.11 kg

8
a 90° c 270° e 90°
b 180° d 180° f 270°

9
a West c South e East
b South d North f East

10
a 28, 32, 36, 40   c 30, 12, 48, 54
b 5, 11, 9, 7   d 2, 3, 6, 8

11
a 20, 30, 40, 60
b 350, 140, 560, 630

12
a 12   b 85   c 5   d 9

Unit 29

1
a 150 b 200 c 360 d 140 e 120
f 350 g 480 h 180 i 150 j 250
k 560 l 180 m 350 n 240 o 300
p 280 q 480 r 640 s 420 t 400

2
a $140 b $180 c $320 d $450

3
a 1 000   b 2 000   c 3 000

4
a 2 000 d 1 000 g 6 000 j 4 000
b 1 000 e 0   h 8 000 k 5 000
c 1 000 f 5 000 i 4 000 l 7 000

5
a 6 090 6 000 reasonable
b 5 037 5 000 reasonable
c 7 967 8 000 reasonable
d 12 111 10 000 unreasonable
e 12 971 13 000 reasonable
f 6 091 8 000 unreasonable

6
a $10   c $44   e $995
b $13   d $26   f $86

7
a 40   c 25   e 35
b 60   d 50   f 210

8–9 Hands on.

10
a 36 m²   b Discussion (6 × 6)
c Hands on.

11 Hands on.

12
a 24 m²   b 12 m²   c 6 m²

13 Hands on.

Answers

Unit 30

1
a 7   e 5r3   i 16   m 12   q 15r1
b 7   f 6r4   j 15   n 12r3   r 14r2
c 8   g 4r2   k 12   o 13r3   s 13r2
d 5   h 5r2   l 18   p 11r3   t 13r2

2
a 16   c 9r3   e 6
b 12   d 8   f 5r3

3 Possible answer:
a 37 ÷ 7 = 5r2   c 37 ÷ 6 = 6r1
b 37 ÷ 5 = 7r2   d 37 ÷ 2 = 18r1

4 a 8   b 7r2   c 5r3   d 16r1

5
a 6414   c 6115   e 2213
b 5326   d 2343

6
a 2315   c 7415   e 3326
b 2152   d 1673

7
a 4546   c 5765   e 2727
b 2332   d 3071

8
a $3000   c $3670   e $1800
b $955   d $5470   f $4515

9 Ford and BMW.

10
a A, E, F, H, J   f D, E, H
b B, C, D, G, I   g A, F, I, J
c C, G   h A, E, F, H, J
d C, D, E, G, H, I   i A, B, C, D, F, G, J
e A, D, E, F, H, I, J

11 Hands on, shapes may be regular or irregular and can be on any orientation.

12 a 1:03   b 1:24   c 1:15   d 12:43

13
a 4 minutes   c 6 minutes
b 8 minutes   d 9 minutes

14
a Loganlea   c Beenleigh
b South Brisbane

15 30 minutes.

16 Hands on.

Unit 31

1
a Witch Bank   c 09/15
b Ms Sava Lotte   d 16

2
a 1st July – 31st July
b 8 (the first entry is the opening balance).
c $285   e $584
d $5000   f $14 215, $14 155

3
a $4590   c $1287   e $2966   g $971
b $4771   d $1873   f $6169   h $6762

4
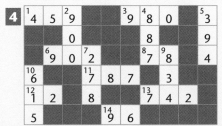

Unit 32

5 Hands on.

6
a Yes   c Yes   e Hands on.
b Hands on.   d Hands on.

7 Hands on.

8

Family size	2	3	4	5	6
Number of families	1	3	5	2	3

9

People per family

(graph: Family size vs Number of families)

10–13 Hands on.

Unit 32

1
a 252   e 1012   i 3360   m 1456
b 384   f 690   j 822   n 537
c 928   g 1524   k 1296   o 2848
d 1235   h 2261   l 1300

2 a 520 g   b 2.1 kg   c 1.2 kg   d 1.8 kg

3
a 4.5   c 4.6   e 31.6   g 52.8
b 2.3   d 5.2   f 41.7   h 73.3

4 The decimal point moved one place to the right.

5
a 25   c 64   e 623   g 825
b 32   d 78   f 714   h 376

6 The decimal point moved two places to the right.

7
a .526   c .986   e 8.56   g 2.53
b .849   d .721   f 9.99   h 7.28

8 The decimal point moved one place to the left.

9
a .856   d .687   g 1.80
b .999   e .325   h 2.75
c .253   f .904   i 3.68

10 The decimal point moved two places to the left.

11
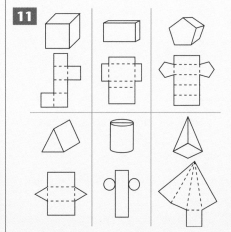

Unit 32

12

Forms a 3D object	a b c e g
Doesn't form a 3D object	d f

13
a 1.37   c 1.52   e 1.23
b 1.48   d 1.69   f 1.95

14–15 Hands on.

16
a 1 m 25 cm, 1.55 m, 2.07 m, 240 cm
b 1.05 m, 1 m 15 cm, 1.55 m, 256 cm
c 1 m 10 cm, 1 m 45 cm, 208 cm, 2.10 m
d 2.04 m, 2 m 15 cm, 2.56 m, 2.65 m
e 0.99 m, 4.99 m, 5 m 01 cm, 5.50 m

17 27 cm

Unit 33

1

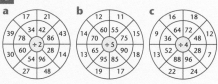

2
a 1   d 8   g 4   j 8
b 1   e 2   h 1   k 1
c 7   f 7   i 4   l 8

3

	Amount	No. of shapes	Amount per share
a	$88	2	$44
b	$96	4	$24
c	$72	8	$9
d	$90	5	$18
e	$80	10	$8
f	$99	9	$11
g	$96	6	$16

4 a 7 km   b $29

5
a 65.56 m   c 38.49 m   e 89.61 m
b 28.05 kg   d 51.78 kg

6
a 5.12 km   c 73.55 km
b 35.03 km   d 38.52 km
e Webb, Jones, Norman and Perkins

7 a D and B   b A and C   c $3009.40

8
a Ball St   c Park St   e Park St
b Newcombe Rd   d Court Ave

9 a A   b K   c L   d G   e S

10–11 Hands on.

12 Grid references
K1, L1, M1, N1, O1, P1, L2, M2, N2

13–15 Hands on.

Unit 34

1 A is the winning card.
a 1596   d 948   g 1936
b 1772   e 1956   h 2622
c 3230   f 3724   i 1184

2 a 596   b 3618   c 3936
d 4228   e 3445   f 6464

Answers

3 Hands on.

4
a 6	**c** 15	**e** 10	**g** 6	**i** 10
b 18	**d** 14	**f** 12	**h** 6	**j** 50

5
a $(5 \times 4) \times 3 = 5 \times (4 \times 3)$
b $4 \times (3 \times 2) = (4 \times 3) \times 2$
c $95 + 5 - 20 \neq 50 + 20 + 5$
d $75 - 30 + 15 = 90 - 30$
e $(24 \div 6) \times 2 = 160 \div 20$
f $6 \times (18 \div 2) = (3 \times 9) \times 2$
g $60 - 15 + 12 \neq 23 + 14 - 5$
h $(7 \times 8) \div 2 \neq 2 \times 4 \times 7$
i $18 - 9 + 18 \neq (35 - 18) + 20$
j $(24 \div 4) \times 2 \neq (30 \div 5) \times 3$

6
a A, C, D, E   c A, C, D   e B
b A, D   d A

7 A, C, D and F
(Only the longer diagonal on shape F divides it into congruent triangles.)

8 a Rhombus   b Rectangle

9 a $120   b $90   c $300
d $240   e $60

10 Hands on. (Subtract costs from previous balances.)

11
	A	B	C	D
8	May 30	Insurance	$ 50	$ 40

12
The Baker family budget
	A	B	C	D
1	*Date*	*Item*	*Cost*	*Balance*
2	Aug 3	Opening balance		$600
3	Aug 4	Groceries	$150	$450
4	Aug 5	Fruit & veg	$500	$400
5	Aug 8	Bills	$100	$300
6	Aug 12	Meat	$ 80	$220
7	Aug 16	Car	$ 60	$160
8	Aug 20	Entertainment	$ 80	$ 80

Unit 35

1
a 615   c 702   e 595   g 804   i 1088
b 912   d 756   f 625   h 896   j 1287

2
a 5   b 4   c 8   d 1   e 6

3
a $375   c $525   e $660
b $532   d $1040

4
a 0.1, 0.2, 0.3, 0.4, 0.5, 0.6
b 0.2, 0.4, 0.6, 0.8, 1.0, 1.2
c 0.3, 0.6, 0.9, 1.2, 1.5, 1.8
d 0.23, 0.25, 0.27, 0.29, 0.31, 0.33
e 1.25, 1.30, 1.35, 1.40, 1.45, 1.50
f 1.67, 1.70, 1.73, 1.76, 1.79, 1.82

5
a 1 ++ 0.5 = 1.5 = 2 = 2.5 = 3 = 3.5 = 4
b 1.5 ++ 0.1 = 1.6 = 1.7 = 1.8 = 1.9 = 2 = 2.1
c 2.3 ++ 0.2 = 2.5 = 2.7 = 2.9 = 3.1 = 3.3 = 3.5
d 1.8 ++ 0.3 = 2.1 = 2.4 = 2.7 = 3 = 3.3 = 3.6
e 2.3 ++ 0.6 = 2.9 = 3.5 = 4.1 = 4.7 = 5.3 = 5.9
f 2.6 ++ 0.8 = 3.4 = 4.2 = 5 = 5.8 = 6.6 = 7.4
g 2.6 ++ 2.2 = 4.8 = 7 = 9.2 = 11.4 = 13.6 = 15.8

6
a 4.2, 6.4, 8.6, 10.8
b 5.5, 10.5, 15.5, 20.5
c 0.6, 0.7, 0.8, 0.9
d 4.4, 4.3, 4.2, 4.1

7–8 Hands on.

9
a flip   b turn

10 Hands on. (Draw identical shapes.)

11
a 7:00   c 11:00   e 9:30
b 8:30   d 7:30   f 6:00

12
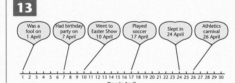
| Sally was born in 1998 | She started walking in 1999 | She started pre-school in 2001 | She started school in 2003 | She started netball in 2006 | She turned 10 in 2008 |

1998 1999 2000 2001 2002 2003 2004 2005 2006 2007 2008 2009

13
| Was a fool on 1 April | Had birthday party on 7 April | Went to Easter Show 10 April | Played soccer 17 April | Slept in 24 April | Athletics carnival 26 April |

1 2 3 4 5 6 7 8 9 10 11 12 13 14 15 16 17 18 19 20 21 22 23 24 25 26 27 28 29 30
Days in April

DIAGNOSTIC REVIEW 4

Part 1
a $60   b $59.74, $59.75   c 25 cents

Part 2
a 130   e 1735   i 13   m about 910
b 210   f 657   j 105
c 448   g 5   k 150
d 592   h 12   l 121

Part 3
a 51.96 m   c 62.22 kg
b 79.92 kg   d 58.95 m

Part 4
a ≈ 1000 c ≈ 400 e approx $3000
b ≈ 400 d ≈ 300

Part 5
a True   b False   c False   d False

Part 6
a 1.2, 2.4, 3.6, 4.8, 6.0, 7.2, 8.4
b 9.5, 9.2, 8.9, 8.6, 8.3, 8.0, 7.7

Part 7
a Riley   d D5   g West
b Taylor   e A3   h North
c A5   f D2   i East

Part 8
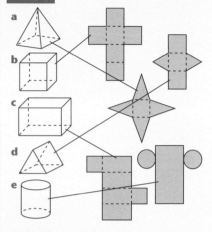
a
b
c
d
e

Part 9
a 12 cm   b 9 cm

Part 10
4 m²

Part 11
2:35, 2:39, 2:45, 2:47, 3:33

Part 12
100 mL